Principles of Biology
A first course

C.V. Brewer
Head of Science, Kingsdale School

C.D. Burrow
Deputy Headmaster, Chosen Hill School

Macmillan Education

© C. V. Brewer and C. D. Burrow 1978

All rights reserved. No part of
this publication may be reproduced
or transmitted, in any form or by
any means without permission

First published 1978

Published by
Macmillan Education Limited
Houndmills Basingstoke Hampshire RG21 2XS
and London
Associated companies in Delhi Dublin
Hong Kong Johannesburg Lagos Melbourne
New York Singapore and Tokyo

Filmset and printed by
BAS Printers Limited
Over Wallop Hampshire

British Library Cataloguing in Publication Data
Brewer, CV
Principles of biology.
1. Biology
I. Title II. Burrow, C D
574   QH308.7
ISBN 0-333-15576-9

Contents

Preface iv

Acknowledgements v

Part 1 The living world
The living world 2

Part 2 The cell
The living cell 6

Part 3 The organism: its diversity
Single-celled animals
   (*Protozoa*) 10
Coelenterates 12
Flatworms 14
Segmented worms 16
Insects 19
Arthropods 26
Molluscs 28
Echinoderms (animals with
   spiny skins) 30
Vertebrates 32
Fish 34
Amphibians 38
Reptiles 40
Birds 42
Mammals 46
Bacteria 50
Viruses 53
Algae 54
Fungi 56
Lichens 59
Mosses and liverworts 60
Ferns 62
Conifers 64
Flower plants 66

Part 4 The organism: in action
Nutrition 72
Transport 84
Energy production 95
Removal of waste 100
The skin 104
Support and movement 106
Plant and animal behaviour 112
Reproduction 123
Growth 133
Heredity 135
Evolution 140

Part 5 The community
The community 144

Index 167

Preface

Principles of Biology has been written for students who have very little knowledge of biology. Biology is first of all about living things and their visual impact: a study of the ways in which organisms function, as individuals or in groups, comes later. An important part of this book, therefore, deals with a variety of plants and animals seen from the outside as whole organisms.

A textbook is mainly a source of information; for the biology textbook, visual information is paramount. The student neither needs nor wants a literary creation. Textbooks are seldom read for pleasure: they are, however, read for reference. As a result, statements that cannot be made concisely have no place in this book. It is intended as a source-book of information to supplement the work of a teacher. The treatment has been made as simple as possible, in order to assist the student who works unaided.

Acknowledgements

The authors and publishers wish to acknowledge the following photograph sources:

Heather Angel pp. 12, 16, 17, 26 centre right, bottom left, bottom right, 28, 29 top, 31 top and centre, 34 bottom, 35 bottom, 36 top, 37, 38 top, 39 centre and right, 41 right and bottom, 48 top centre, 54, 62 top right, centre right, 63 top left, centre left, top right, 64 top left, top right
Barnabys pp. 23, 24 top left, 27 top right, 38 bottom left, 39 left, 40, 41 top left, top centre, 49 top left, 53 left, 56, 60 top, 62 bottom, 64 bottom, 90, 124 top
Brian Bracegirdle (Photographer for science) pp. 115 top right
Elizabeth Brewer pp. 158, 160, 161 bottom right, 162
British Museum (Natural History) pp. 48 bottom centre, 142
Cambridge Scientific Instruments p. 81
Camera Press pp. 49 bottom left, 65 right, 133 top left and right, 141
J. Allen Cash pp. 18, 24 top right, 26 top, 27 top left, 35 top, 42, 43, 46 top and top insert, 48 top right, 49 bottom right, 59, 151 top left, centre right
Bruce Coleman Ltd p. 49 top right
(Des & Jen Bartlett) pp. 24 bottom left, 36 bottom left
(S. C. Bisserot) pp. 58 left, 141
(Jane Burton) pp. 24 bottom right, 29 bottom, 30 top, 36 bottom right, 49 top centre, 62 top left
(Oxford Scientific Films) pp. 26 centre left, 30 bottom left, centre right
Countryside Commission p. 166
Gene Cox pp. 6, 50, 55 top left, 58 right, 61 top, top centre left, top centre right, 63 centre right, bottom left, bottom right, 86 bottom, 99, 120
Alan W. Curtis p. 48 centre left
J. E. Downward p. 123 bottom left
Hermann Eisenbeiss cover
E. A. Ellis p. 124 bottom
Forestry Commission p. 151 top right, bottom right
Glasgow University p. 114
Horniman Museum pp. 38 bottom right, 48 centre right, 77, 78, 110
Institute of Geological Sciences p. 140 top
IBM p. 115 bottom
Keystone Press Agency Ltd pp. 46 bottom left, 49 bottom right
Kings College Hospital (Geoffrey Chamberlain) p. 48 bottom left
Dr G. F. Leedale (University of Leeds) p. 155
Meridan Airmaps p. 161 top right
Micro Colour International (Charles Cook) p. 14 bottom
Novosti p. 140 bottom
Popperfoto pp. 11, 157 top
Dr M. C. F. Proctor pp. 60 bottom left, bottom right, 61 bottom centre left, bottom
RTHPL pp. 46 bottom right, 111
Graham Reed pp. 2, 76, 133 bottom
Rothampsted Experimental Station pp. 74, 161 top left
William Rowntree p. 31 bottom
John Sankey p. 161 centre
Science Museum pp. 55 bottom right, 65 top left
Tony Seddon p. 151 centre left
Shell Photographic Service pp. 21, 27 centre and bottom
Soil Survey of England and Wales p. 153
F. Thompson p. 152
C. James Webb pp. 25, 48 top left, bottom right, 86 top, 104, 115 top left, 118, 123 top left and right, 132 136
Dr D. P. Wilson p. 157 centre and bottom

The publishers have made every effort to trace the copyright holders, but if they have inadvertently overlooked any, they will be pleased to make the necessary arrangement at the first opportunity.

Part 1
The living world

The living world

How many articles in this photograph are derived from living organisms?

Biology is the study of life. There are many reasons why biology is important to man. Most of his needs for food and shelter are supplied by living organisms. Diseases are caused by living organisms.
Many plants and animals are 'pests' – they compete with man for food and material.

Biology began with simple observations of the living world and continued with collections of specimens of plants and animals. As the numbers of known organisms increased, it became important to catalogue and classify them.

Although a knowledge of human anatomy dates back to the ancient Greeks, the study of the internal parts of plants and animals did not really begin until the seventeenth century, following the invention of the microscope.

In modern biology, living organisms are studied at all levels – at one extreme, the biochemist investigates the chemicals of living organisms; at the other, the ecologist studies the relationships of plants and animals in their natural environments.

The requirements of the living world

Living organisms need:

light
a supply of oxygen and carbon dioxide
the presence of water
a suitable temperature.

The *sun* is the only source of energy for the living world. This energy reaches the earth in the form of light and heat. The light energy is used by green plants to make energy-rich materials (food). The heat energy makes sure that the temperature of most of the earth lies within the range over which life can exist – not so cold that water is in solid form only (ice) and not so hot that most water evaporates.

The *atmosphere* covering the earth is a store for essential gases – carbon dioxide, oxygen and water vapour. The amount of carbon dioxide is quite small and its turnover is rapid. There is much more oxygen, and its turnover is slower. The atmosphere gives protective cover, absorbing harmful rays (especially ultra-violet); it also restricts loss of heat and gives the land surface relatively unchanging temperatures. Air is an important medium for dispersal – seeds, spores, flying animals and even organisms such as spiders, light enough to be carried by the wind.

Water is an important heat store. It takes up and loses heat slowly. Even more than the atmosphere, the oceans cushion the earth against

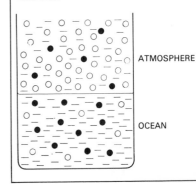

Diagram showing that atmosphere and ocean differ only in their proportions of water⁻, oxygen°, and carbon dioxide•

Principles of Biology

temperature changes. Water is a very stable medium. It is much denser than air and gives better support. It holds oxygen, carbon dioxide and salts in solution: it is a liquid closely resembling body fluids. Because of this, it provides the best possible home for living organisms.

The materials of the living world

All living organisms are composed of the same materials – water and carbon compounds. Water is important because other substances can dissolve in it; a large part of every living organism is water, and the essential chemical processes of life take place in solution.

Carbon compounds are created by green plants using energy from the sun. The energy is then locked into these carbon compounds which can be used for tissue-building or can be broken down to release their energy in movement or as heat. Carbon compounds provide the means for transferring the sun's energy from one living organism to another; the carbon compounds of one organism become the food of the next organism in the chain.

The basic bulk food substances are the carbon compounds carbohydrates, fats and proteins.

The origins of the living world

The present-day appearance of the earth, the existence of soil, and the composition of the atmosphere are almost certainly the result of the influence of living organisms over millions of years. Life has existed on earth for at least 3,000 million years. Before this, after the earth had cooled, it consisted of oceans and bare rock. The atmosphere was very different: apparently there was

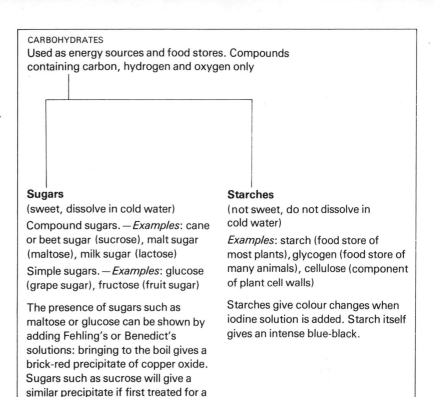

CARBOHYDRATES
Used as energy sources and food stores. Compounds containing carbon, hydrogen and oxygen only

Sugars
(sweet, dissolve in cold water)
Compound sugars. — *Examples*: cane or beet sugar (sucrose), malt sugar (maltose), milk sugar (lactose)
Simple sugars. — *Examples*: glucose (grape sugar), fructose (fruit sugar)

The presence of sugars such as maltose or glucose can be shown by adding Fehling's or Benedict's solutions: bringing to the boil gives a brick-red precipitate of copper oxide. Sugars such as sucrose will give a similar precipitate if first treated for a few minutes with dilute hydrochloric acid (the acid is then neutralised or poured away).

Starches
(not sweet, do not dissolve in cold water)
Examples: starch (food store of most plants), glycogen (food store of many animals), cellulose (component of plant cell walls)

Starches give colour changes when iodine solution is added. Starch itself gives an intense blue-black.

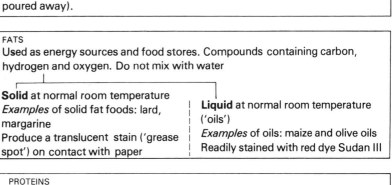

FATS
Used as energy sources and food stores. Compounds containing carbon, hydrogen and oxygen. Do not mix with water

Solid at normal room temperature
Examples of solid fat foods: lard, margarine
Produce a translucent stain ('grease spot') on contact with paper

Liquid at normal room temperature ('oils')
Examples of oils: maize and olive oils
Readily stained with red dye Sudan III

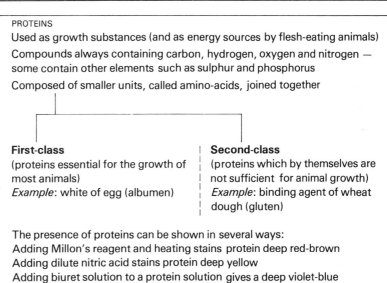

PROTEINS
Used as growth substances (and as energy sources by flesh-eating animals)
Compounds always containing carbon, hydrogen, oxygen and nitrogen — some contain other elements such as sulphur and phosphorus
Composed of smaller units, called amino-acids, joined together

First-class
(proteins essential for the growth of most animals)
Example: white of egg (albumen)

Second-class
(proteins which by themselves are not sufficient for animal growth)
Example: binding agent of wheat dough (gluten)

The presence of proteins can be shown in several ways:
Adding Millon's reagent and heating stains protein deep red-brown
Adding dilute nitric acid stains protein deep yellow
Adding biuret solution to a protein solution gives a deep violet-blue

no oxygen, but gases such as methane and ammonia were present, and much more carbon dioxide.

It is known that, after thunderstorms, nitrates are formed from the nitrogen of the air. In similar extreme conditions, mixtures of gases containing methane and ammonia produce amino-acids. Where there is no oxygen to break them down, amino-acids and other similar compounds formed by accident gradually increase in concentration. When this process continues for millions of years, it becomes inevitable that combinations of compounds appear which show the properties of life. Some of these early organisms must have synthesized their own food using carbon dioxide and the sun's energy, giving off oxygen as a waste product.

Fertile soils are formed when bare rock weathers. The action of wind, rain, running water, ice and extreme temperature-changes gradually splits the rock into fine particles. This provides the basis of soil but, without living organisms, soils are not stable or fertile – the small particles are blown or washed away. Plants help in the breaking-down process by the action of their growing roots, and they add to the soil fertility with their dead remains. Growing plants also prevent erosion by wind and water in two ways – they give surface cover and their roots bind the soil particles together.

The creation of a living world, then, seems largely a matter of chance, but it also has a certain inevitability. For this reason, there must surely be life elsewhere in the universe.

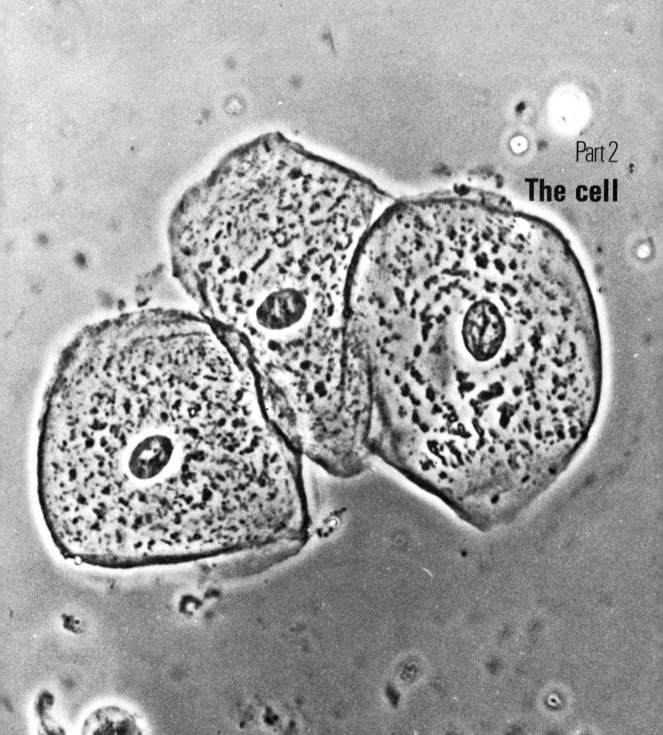

Part 2

The cell

The living cell

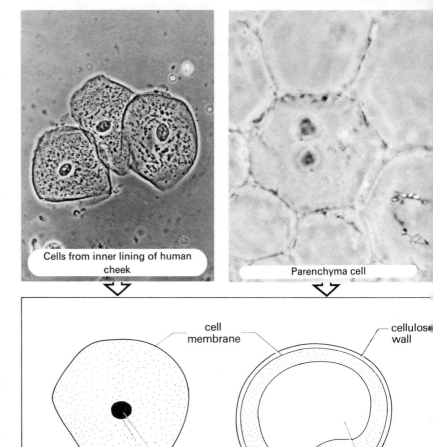

Cells from inner lining of human cheek

Parenchyma cell

All life processes take place within cells. The properties of living things, then, are properties of cells.

Cells can:

Take in materials from outside for use as food.
This is called *nutrition*.

Change food materials (particularly amino-acids) into their own living material.
This leads to an increase in size called *growth*.

Break down some food materials (particularly sugars), changing the energy they contain into heat energy, energy for movement, etc.
This is *respiration*.

Remove the waste materials formed by their activities.
This is *excretion*.

Produce living, but quite separate, likenesses of themselves.
This is *reproduction*.

React to changes around them.
This is *irritability*. A common reaction is *movement* of some kind.

Cells are the basic units of life. Protozoa, bacteria and some algae consist of a single cell only. Other living organisms are many-celled.
Size is critical for cells. Their narrowest dimension is seldom less than 0.01 mm or more than 0.1 mm. Viruses are very much smaller but they are incomplete as living organisms and depend on living cells to supply the materials they lack. Egg cells are usually very much larger, but are not active.

Cells collect on the blunt end of a nail file when this is gently scraped against the inside of the cheek. These animal cells can be examined in a drop of water under a microscope. Plant cells can be seen in a thin shaving from the cut surface of a stem, also examined in a drop of water.

Each cell has three main parts:

An outer membrane forming the cell boundary.
The cytoplasm, a transparent, rather watery jelly, sometimes containing granules. Cytoplasm often shows flowing movements.
The nucleus, less watery and lying in the cytoplasm. It has its own membrane and is usually rounded or disc-shaped.

A plant cell has two additional features. It is enclosed in a firm wall of cellulose. This gives it more support and makes the boundaries of plant cells more clearly visible. Plant cells also have a central space (vacuole) filled with sap. Cell sap is a watery solution of sugars and salts.

Principles of Biology

Cells often contain other materials – stored food such as fat or starch, pigments such as haemoglobin (in animals) and chlorophyll (in plants), and insoluble waste materials.

A cell is a factory in miniature. It takes in food which is a source both of energy and of raw materials. Just as industry may use oil to generate electricity or as a raw material for making plastics, paints and perfumes, so a gland cell in the pancreas will use soluble food to provide its energy (by respiration) and to make insulin.

All cells carry out many manufacturing processes as well as releasing energy from food materials. The nucleus is the control centre and stores all the information needed to maintain the cell's activities and to construct new cells. Because of this, cells cannot be smaller than a certain size.

Similarly, they must not be too large. Materials such as oxygen for respiration are absorbed through the cell surface: they enter by diffusion. As a cell grows larger, its surface area increases less than its volume. A large cell obtains essential materials or removes its waste less efficiently: it either divides or dies.

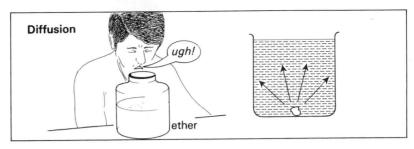

DIFFUSION is the movement of molecules from higher to lower concentrations.

Osmosis
Osmosis is a special type of diffusion: it applies to water molecules and thin barriers (called semi-permeable membranes) that allow only certain molecules to pass through.

Osmosis occurs when a red blood cell is placed in different solutions.
OSMOSIS is the movement of water from a region of high water concentration to a region of lower water concentration through a semi-permeable membrane.

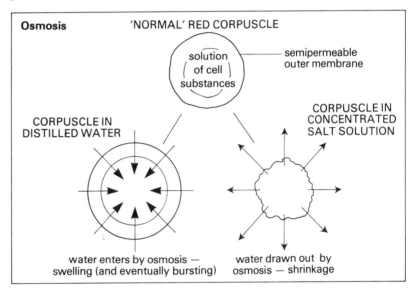

Diffusion
Molecules of substances are in constant motion. They move in all directions – their movement is random. In gases, molecules move at high speed; in liquids they move more slowly; in solids, they vibrate.

Particles spread out into all the space available to them. They spread from regions of high concentration to regions of low concentration. This movement is called diffusion: it applies to gases in the atmosphere and to materials in solution such as oxygen and sugar.

The living cell 7

Part 3
The organism: its diversity

Single-celled animals (Protozoa)

These are extremely small, though the largest can be seen without a microscope. Unlike most animals, the body has one cell only, and this is able to live entirely by itself, controlled by its nucleus.

Protozoa are found in most places – in freshwater ponds and streams, the sea and the soil; some live even within the bodies of other organisms, where they may be parasites and cause disease (examples include malaria and sleeping sickness).

Oxygen needed for respiration is usually dissolved in the surrounding water; it diffuses inwards all over the surface of the animal. Carbon dioxide diffuses out in a similar way, together with other waste materials such as ammonia.

Single-celled animals grow only by increasing the size of their cells. Food is used to make more protoplasm until the cell becomes so large that it divides, usually into two half-sized cells. Division into two cells is called binary fission.

A Flagellum beats to pull animal through the water.
Eye spot helps animal to detect suitable conditions of light.
Chloroplasts make food by photosynthesis.
(*Euglena* lacks cellulose wall and sap vacuole typical of most plants.)

B Cilia are short threads of protoplasm which beat against the water and drive animal along in a spiral. Special cilia in the cell gullet beat to carry food particles (bacteria, etc.) in a water current towards mouth.

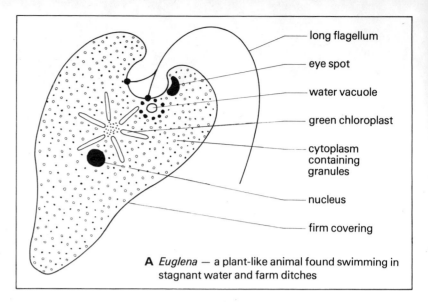

A *Euglena* — a plant-like animal found swimming in stagnant water and farm ditches

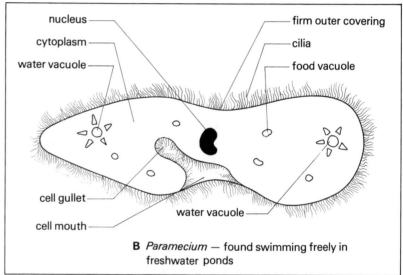

B *Paramecium* — found swimming freely in freshwater ponds

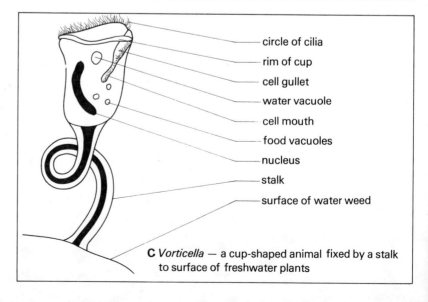

C *Vorticella* — a cup-shaped animal fixed by a stalk to surface of freshwater plants

Principles of Biology

C Cilia are used to trap bacteria for food; not for movement. Stalk can contract to draw animal closer to its attachment.

D Pseudopodium ('false foot') provides a slow, flowing movement – it becomes larger and longer as protoplasm slowly flows into it from other parts of the cell: also surrounds and engulfs (ingests) food.

Food vacuoles contain food particles (bacteria, etc.) in a drop of water, digested by enzymes secreted by *Amoeba* into vacuole.

Water vacuoles are used by freshwater animals to prevent flooding; vacuole gradually fills with water which continually passes into animal by osmosis: when full, vacuole bursts, pumping its water out into surroundings.

All single-celled animals show some sensitivity. *Amoeba* flows away from bright light, contact with sharp points (needles) and unpleasant chemicals (acids, etc.). *Amoeba* senses the presence of nearby food: it forms a pseudopodium, flows towards the food, surrounds it and takes it into a food vacuole with a drop of water (ingestion). After being killed and digested by enzymes, useful materials are absorbed into the cytoplasm. Undigested remains are forced out from the cell as *Amoeba* flows away (egestion).

Amoeba can reproduce by itself – this is asexual reproduction. The animal stops moving and feeding; first the nucleus and then the whole cell divides, forming two small cells. Each begins moving, feeding and growing.

Some *Amoebae* stop moving and secrete a hard covering (cyst) around themselves. Inside this, they can survive drought or freezing conditions. They may even be dispersed to new ponds, blown by the wind or carried on the feet of birds. Sometimes the cell inside the cyst may divide many times, so that when the cyst eventually breaks open a large number of very small *Amoebae* emerge, instead of one only.

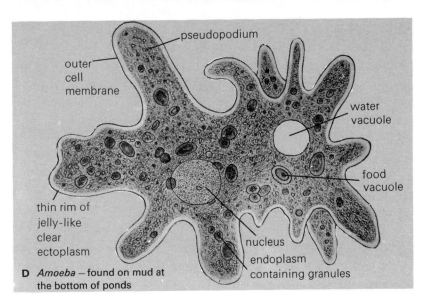

D *Amoeba* – found on mud at the bottom of ponds

Single-celled animals (Protozoa)

Coelenterates

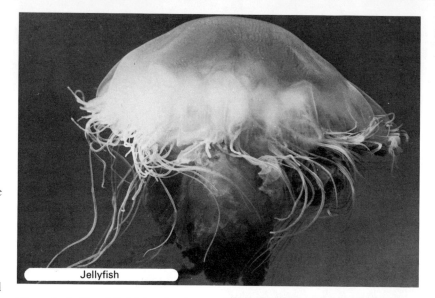
Jellyfish

Coelenterates are aquatic animals most of which live in the sea. The animal consists of a body wall made of two layers of cells separated by a thin film of jelly, surrounding a cavity which serves as a gut. There are no organs. There is a single opening, serving both as a mouth and as an exit for undigested remains. The mouth is surrounded by a ring of tentacles. The body may be cylindrical, with mouth and tentacles at one end and a base at the other, attached to a rock or water plant (sea anemone). It may be flattened from top to bottom and turned the other way up like an inverted saucer (jellyfish).

The parts of a coelenterate are radially arranged, like a buttercup flower.

Coelenterates either stay in one place (sea anemone) or drift about in the surface waters of the ocean (jellyfish). All are carnivores, fishing for small animals by means of tentacles which are covered with batteries of stinging cells. Coelenterates absorb oxygen in solution over their entire surface, carbon dioxide and other waste materials diffusing in solution into the water. Many coelenterates are colonial; this means that their bodies are actually joined together. Some of these produce a stone-like external supporting skeleton (coral) which forms the reefs and atolls of the Pacific and Indian Oceans and the Caribbean Sea.

Solitary animals
Anemones are found in rock pools between low and high tide marks.

The cavity inside a jellyfish is reduced to a system of narrow canals by the very thick jelly layer.

Hydra
The hydra is found in ponds and slow-moving streams. It is attached to water weeds or may be upside-down in the water, suspended from the surface film. Some sorts are green. The colour is due to the presence of single-celled algae which live inside the cells of the hydra. This is an example of symbiosis, a relationship which benefits both organisms. In the daytime the algae photosynthesise,

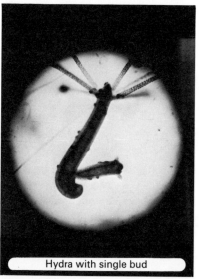
Hydra with single bud

Sea anemone feeding on a fish

Principles of Biology

producing oxygen and using up carbon dioxide. The cells of the hydra respire, producing carbon dioxide and using up oxygen.

The hydra feeds on tiny animals such as water fleas. The stinging cells are triggered by contact and eject their threads. Some stick to

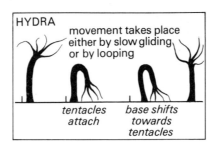

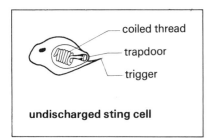

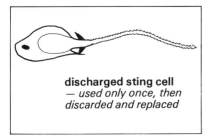

the prey. Others harpoon it, injecting a paralysing poison at the same time. The water flea is held fast to the tentacle. Other tentacles move in and help draw the water flea into the mouth. After digestion has taken place the undigested remains are pushed out through the mouth.

When fishing for food, the hydra is long and thin. Its tentacles form long threads of fishing lines. This is because the cells lining the gut cavity and the hollow tentacles have muscle strands running circularwise, and they are contracted. When disturbed, the hydra and its tentacles become short and fat. The cells in the outer cell-layer have muscle strands which run lengthwise and these are now contracted. Hydra is able to move about either by sliding on its base or by looping.

When food is plentiful the hydra grows buds which are outpushings of the body wall. These form head and tentacles and eventually drop away. Towards the mouth end of the body, unspecialised cells in the outer layer divide many times to form sperm cells. Further towards the base, similar cells also divide. One of them enlarges at the expense of the rest and becomes an egg cell. It is fertilised while still in the body wall by a sperm cell from another hydra, and develops into a ball of cells within a horny case. It is then shed and falls to the bottom of the pond where, when conditions are suitable, it grows into a new hydra.

Coelenterates 13

Flatworms

Flatworms are many-celled animals with very thin bodies, often shaped like a leaf. Many species of flatworms live as parasites inside the body of other animals.

A PARASITE is a living organism which attaches itself to, and takes its food from, a different living organism, called a host. The host suffers but is not normally killed by the parasite.

Flatworms have very complicated reproductive systems and usually each animal is both male and female. As in coelenterates (sea anemones, etc.), the digestive system has only one opening, though in some parasitic flatworms there is no digestive system at all.

Planarians

These are free-living flatworms, found in wet places. Many live in ponds and streams, but only if the water is unpolluted; they hide under stones during the day. Their bodies are variously-coloured – yellow, orange, grey, black. The head end is poorly developed and gradually merges with the body. The pair of eyes at the front are sensitive to changes in light intensity. Planarians emerge to feed in the dark. Most feed on the dead remains of other water animals. The throat is pushed out like a sac through the mouth and closes over the food, which is then sucked in. When they are starved, planarians will digest their own organs and 'degrow' in a definite sequence.

Planarians swim in a characteristic zig-zag fashion by muscle contraction. They can also glide over surfaces in a film of slime (mucus)

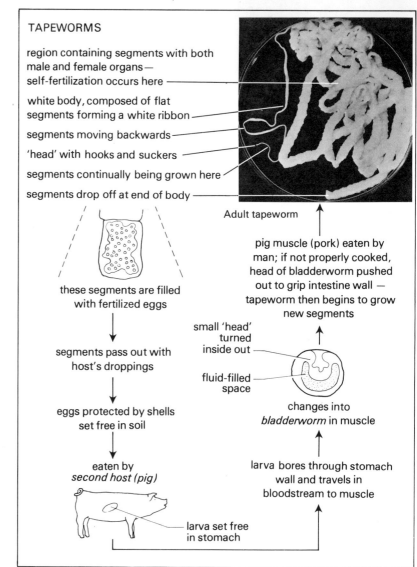

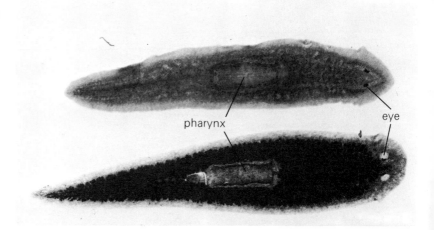

Planarians

Principles of Biology

which they secrete. They push themselves along by the beating of cilia – short threads of protoplasm that cover the body.

Planarians normally mate, and cross-fertilization occurs before the eggs are laid.

Parasitic flatworms

Tapeworms
Example: **pork tapeworm**, found in small intestine of man.

Food digested by host absorbed over entire body surface. Surface of parasite protected from digestive juices of host by tough cuticle. Enormous numbers of fertilized eggs are produced, because chance of infecting new host is extremely slight.

Flukes
Some (blood flukes) live inside blood vessels. The **liver fluke** lives in the bile duct of sheep, attached by suckers. It feeds on the tissues of the bile duct and on blood. This causes fatal 'liver rot'. The adult flukes are protected by a tough, spiny cuticle.

After mating and cross-fertilization, many eggs are laid. They leave the sheep with its droppings. The life-cycle requires a *water snail* as a *second host*. Because of this, the disease most often affects sheep living in damp pastures. The parasite produces enormous numbers of offspring since it is extremely unlikely that sheep would become infected otherwise.

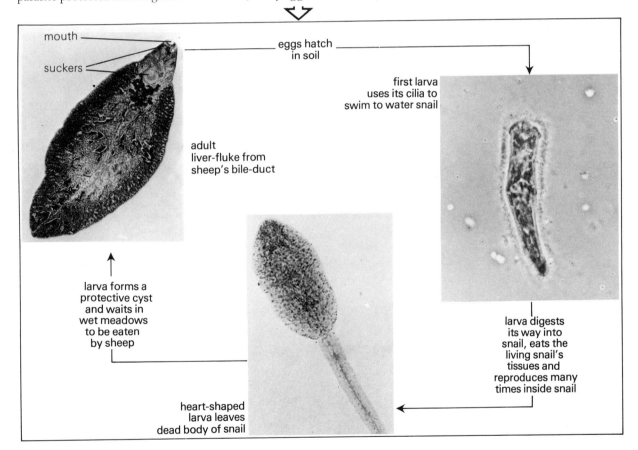

mouth
suckers
adult liver-fluke from sheep's bile-duct
eggs hatch in soil
first larva uses its cilia to swim to water snail
larva digests its way into snail, eats the living snail's tissues and reproduces many times inside snail
heart-shaped larva leaves dead body of snail
larva forms a protective cyst and waits in wet meadows to be eaten by sheep

Flatworms 15

Segmented worms

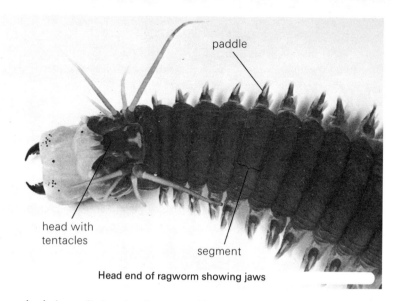

Head end of ragworm showing jaws

These are sometimes called the 'true worms'. Their bodies are clearly divided into a series of similar rings; this arrangement is continued inside the body. Each segment may contain, for example, excretory and reproductive organs in addition to muscles, blood vessels and nerves. The digestive system extends down the centre of the body like a tube within a tube; there is an opening (mouth) at the front and a second opening (anus) at the hind end. The body surface is soft and moist and (except in leeches) has projecting bristles. There are no truly hard parts; the space between the two tubes (between outer body wall and digestive system) is filled with fluid. The pressure of this fluid helps to support the body. There are no limbs. Glands in the skin secrete a slimy mucus. Oxygen, for respiration, dissolves in the mucus and diffuses through the thin outer covering into small blood vessels.

Ragworms, lugworms and tubeworms
These are all marine worms, found mainly on the sea shore. Like earthworms, they are used as bait for fishing. The body bears many bristles, usually carried on special paddle-like sideways extensions of the body. Each worm is either male or female, and the head region is better developed than in earthworms and leeches.

Ragworms live near the low-tide mark; they lie buried in sand or hide under stones by day. The cream or green body has a distinct head with eyes and several pairs of sensory tentacles, which assist in finding and catching small animals for food. The throat has horny toothed jaws which can be pushed out through the mouth during feeding. Bundles of bristles stand out from a pair of paddles at the sides of each segment. These paddles help the worm to swim or crawl, using wriggling movements. Fertilization occurs in the sea after sperm and eggs are shed into the water; the eggs hatch into free-swimming larvae which later settle and become adult.

Lugworms are the earthworms of sandy shores. They burrow into mud and sand near low water. Coiled casts can be seen as the tide goes out: they result from the lugworms feeding on mud and digesting the small organisms it contains. The head and paddles are not as distinct as in ragworms. Many of the segments have tufts of gills.

There is a great variety of **tubeworms**. Some secrete a chalky tube, others live in tubes made of sand which the animal glues together. The tentacles on the head may emerge to catch food; at other

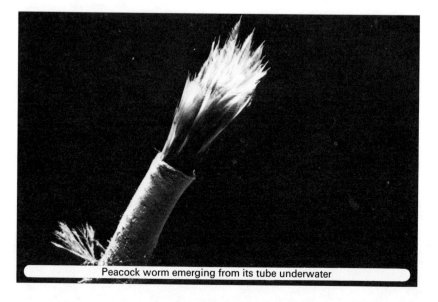

Peacock worm emerging from its tube underwater

times, and when the tide is out, the worm remains hidden and protected in its tube.

Leeches

Most leeches are blood-sucking parasites which live in damp places or in fresh water. They attach themselves for short periods to the bodies of other animals by a sucker at the front and rear end. The head is indistinct. The mouth has three jaws which puncture the skin, and the throat draws out the blood. Saliva is pumped into the wound to prevent the blood clotting. The body is very elastic and the digestive system has many storage pouches, so that leeches can live for long periods without a meal, swollen with several times their own weight of blood.

Leeches make looping movements, using one sucker as anchor and then sliding the rest of the body forwards. Some can swim by gentle undulations of the body. Leeches are most active at night and are strongly attracted by the vibrations of nearby animals.

Each leech has male and female organs. Sperm is exchanged at mating, so that there is cross-fertilization. As in earthworms, cocoons are formed from which young leeches emerge later.

In the past, leeches were used by doctors to bleed patients in an attempt to treat illness.

Earthworms

A common British earthworm (*Lumbricus*) may reach a length of 30 cm, but giant earthworms of South America and Australia are more than 2 m long. Earthworms have cylindrical bodies, tapering at each end. There may be well over 100 segments, but no head. In mature worms, the body is swollen to form the 'saddle' about one-quarter of the way back. The upper surface is darker (red-brown) and the lower surface feels rough because of its pairs of protruding bristles. There are no large sense organs such as eyes but the skin has clusters of cells sensitive to touch, temperature changes and chemical substances.

Each worm has male and female reproductive organs, with openings on the underside about midway between mouth and saddle. Nevertheless, there is always cross-fertilization following the mating process, which usually occurs on warm, damp nights at the soil surface. The worms lie alongside each other, pointing in opposite directions with their undersurfaces touching. They are held together

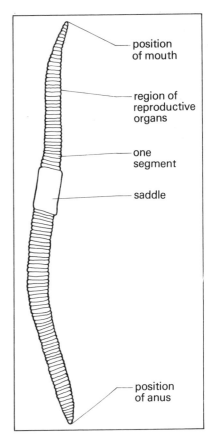

by a mucus tube secreted by the saddle, and by bristles thrust into the other worm. Sperm is then exchanged. The worms separate and each secretes a lemon-shaped cocoon; eggs and sperm are passed in and fertilization takes place. Later, one young worm emerges from each sealed cocoon.

Earthworms burrow by day. They avoid light and will surface only when the soil is flooded. Their burrows may be more than 2 m deep, with the main upper part nearly vertical but winding below into a larger chamber in which one or more worms live. When the soil is soft, the front end of the worm makes a wedge in the soil and the body swells to push the soil aside. Heavy soil is eaten, drawn in by the muscular front end of the digestive system. Organic matter is digested and the soil finally pushed out through the anus to form worm casts at the entrance to the burrow.

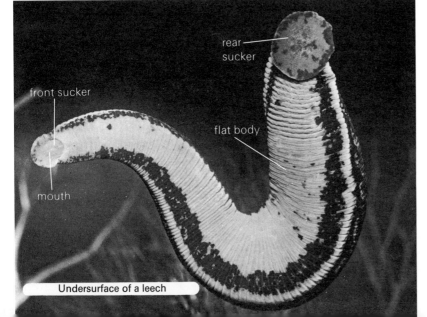

Undersurface of a leech

Segmented worms 17

Earthworms mating

Mucus is used to line the burrow and the entrance is plugged with leaves.

The body wall has two sets of muscles. When the circular muscles contract, they squeeze on the fluid in the body space; this makes the worm long and thin and pushes the front end forward. The bristles are then pushed out to grip the soil for anchorage. Next the longitudinal muscles contract to make the body short and fat, and the rear end is drawn forward segment by segment.

Earthworms are abundant in rich, moist, fertile soils, where their activities help to maintain soil fertility. There may be as many as 7.5 million worms per hectare. Worms are absent from poor soils. Their tunnels bring air into the soil, assist drainage and loosen the soil so that roots can grow more easily. Soil is ground in the worm's gizzard and is cast near the surface; this action mixes and ploughs the soil and provides fine soil suitable for seed germination. Dead remains are dragged down by worms and these break down quickly into plant food; also, the dead bodies and excretions of the worms themselves provide useful manure.

Worms, when damaged, for example by a spade, can grow new hind ends if the front end is undamaged.

The red freshwater worm *Tubifex* is related to earthworms. It is eaten by a variety of animals, both in the wild and in aquaria.

Insects

Insects are one of the largest and most important groups of animals. Although there are about a million species so far described, they are all remarkably alike in their basic features. It is difficult to mistake an adult insect for anything else. The key features which may readily be seen and which should be remembered are:

The possession of an exoskeleton. This is a tough 'skin' which fits the insects like a suit of armour. It is protective; it forms a rigid framework for the attachment of muscles, like the bony framework of a mammal; it is waterproof and therefore reduces water loss by evaporation.

The division of the body into three distinct regions, the head, the thorax and the abdomen. The thorax and abdomen are further divided into segments.

The head carries a pair of compound eyes, a pair of feelers (antennae) and jointed mouth parts.

Each of the three sections of the thorax has a pair of legs. Sections two and three also each carry a pair of wings.

All insects are small. Although some butterflies have a wingspread of several centimetres, no part of an insect is ever more than a few millimetres from the surface.

These are general features. There are, of course, exceptions. Ants have wings only for a part of the adult's life span. Flies have only one pair of wings. Fleas have no wings at all. Other features, less obvious but equally important, are:

Growth occurs in steps. The exoskeleton is dead and does not grow. The exoskeleton is shed (moulted) and a period of rapid growth follows, before the new, soft skin hardens; growth then stops until the next moult. The period in the life of the insect, from one moult to the next, is called an instar.

The young form of an insect is often quite different from the adult. The larva, as it is called, usually has different eating habits and lives in a different place. In some insects (butterfly, bee, housefly) it is so different that a special instar, the pupa, occurs during which complete reorganisation takes place.

Air holes (spiracles) along the sides of an insect's body open into a system of branching tubes (tracheae) which form a network of tiny airways running all over the body like the blood capillaries in a mammal. Although the larger tubes can be ventilated by breathing movements, oxygen can pass along the smaller tubes only by diffusion. Such a system is efficient only over very short distances.

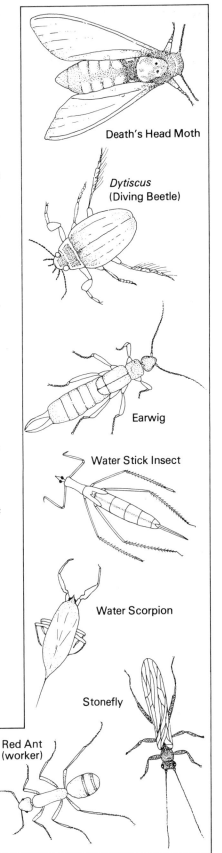

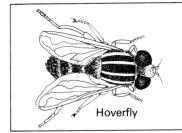

Some common insects (not to scale)

The honey bee

Honey bees are social insects. The swarm consists of individuals which are all developed from eggs laid by a single queen bee. The nest, or hive, is made of sheets of wax. Each sheet consists of two layers of six-sided cells arranged back to back and suspended vertically to form a comb. Three sorts of cell are present. Brood cells are used either as food stores, or they contain eggs which develop into worker bees. Drone cells contain eggs which develop into drones (males). Thimble-shaped cells are also present. These are used for rearing new queens.

Worker bees are female. The eggs hatch after three days. At first the larva receives 'royal jelly', fed to it by workers who produce it from special glands. Later, this is replaced by a diet of honey and pollen. After about six days the larva is sealed in and changes into a pupa. Eight days later the adult worker bee emerges from the pupa and bites its way out.

The adult spends about three days on cleaning duties. After this she feeds the older larvae. At about two weeks she is able to feed the young larvae with royal jelly. She also helps with comb building. From about three weeks after pupation she will spend the rest of her life gathering nectar and pollen and about five weeks later she dies.

Drones develop from unfertilized eggs, laid mostly by the queen but occasionally by the workers. Queens develop from larvae which are fed entirely on royal jelly. The queen's task is simply to lay eggs. The queen cells are built in early summer when the colony is increasing in size.

Just before the first of the pupae in the queen cells is due to appear, the old queen leads some of the colony away to set up a nest elsewhere. The first new queen to emerge from pupation destroys all the others by stinging them in their cells. She produces a secretion from mouth glands which is spread over her body as she grooms herself. As other bees touch her they receive the secretion which has the effect of keeping all the bees together as a colony. Within a few days she mates on the wing with one or more of the drones, during what is called the nuptial flight. From then on she will lay eggs, sometimes as many as 2 000 a day. The queen and her worker bees pass the winter in the nest, feeding on stored food.

Bees are known to be able to communicate with their fellows, by means of a special dance, which gives information about the distance and direction of abundant sources of nectar.

A honey comb

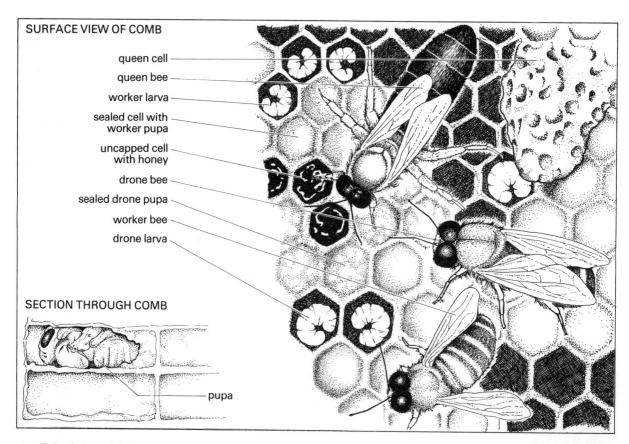

The locust

Locusts are found all over North Africa, in the desert regions of the Middle East, and in Pakistan. The locust feeds on green vegetation, which it cuts up with its powerful jaws. Mating occurs as soon as adults emerge from the final moult. The female digs a hole in damp sand and lays about sixty eggs. These are covered with a frothy substance which hardens to form an egg pod. Young locusts are called hoppers. They are unable to fly since the full-size wings only appear after the final moult. When locusts swarm they may cover an enormous area and cause devastation to any crops which lie in their path. The time from egg-laying until the adult appears is about five or six weeks.

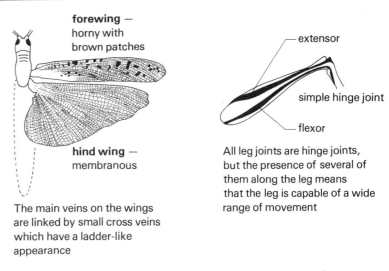

The main veins on the wings are linked by small cross veins which have a ladder-like appearance

All leg joints are hinge joints, but the presence of several of them along the leg means that the leg is capable of a wide range of movement

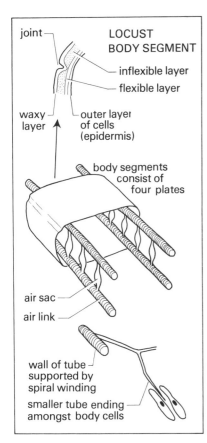

Adult locust

Magnified eye surface

Third Instar Hopper (right), Fifth Instar Hopper (far right)

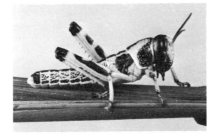

Insects 21

The Large White butterfly

The butterflies appear in late spring. They mate and the female lays her eggs on the underside of cabbage leaves. The caterpillars hatch about a week later. They eat their eggshells and then begin to eat the leaf, using their pincer-like mouth parts to tear off pieces of leaf. The caterpillar grows rapidly, shedding its skin about four times.

After about a month the caterpillar leaves the food plant and seeks a dry, sheltered place. It spins a silk pad to which it attaches itself. It then moults to reveal the pupa. No activity is visible from the outside but drastic changes are going on inside. About three weeks later the pupa case splits and the adult butterfly emerges. By this time, it is late summer. This new generation also mates and lays eggs. The eggs hatch into caterpillars which will not complete their development until the following spring. They pass the winter in the pupa stage. Many will not survive.

The butterfly feeds on the nectar from flowers which is obtained by means of mouth parts which form a tube (proboscis). It is rolled up underneath the head when not being used.

The housefly

The housefly is a two-winged insect. In place of the second pair of wings it has a pair of drumstick-shaped rods which vibrate in flight and act as stabilizers. The fly eats sweet, sugary food. It pours saliva on to the food and sucks up the dissolved material with its tubular mouth parts (proboscis). Eggs are laid during the summer in any decomposing organic matter (manure heaps, dead animals, dustbin refuse). Within a day the small white larva (maggot) hatches.

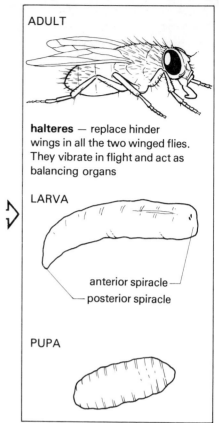

It moults twice during its growth which is completed within a week. It then moves to a drier place and changes into a barrel-shaped pupa which darkens in colour. Four days later, the pupa case splits and the adult fly emerges.

The habits of the housefly, feeding on sweet sugary foods and laying eggs in decomposing material, together with its hairy body, make it a very effective carrier of bacteria and disease. Most flies die off in the winter, but a few survive in sheltered places until the following spring.

The gnat

Gnats, like houseflies, are two-winged insects. Eggs are laid in stagnant water. A single batch is glued together to form a raft, floating on the surface. The larva feeds on scraps of plant and animal material. It takes in air from the

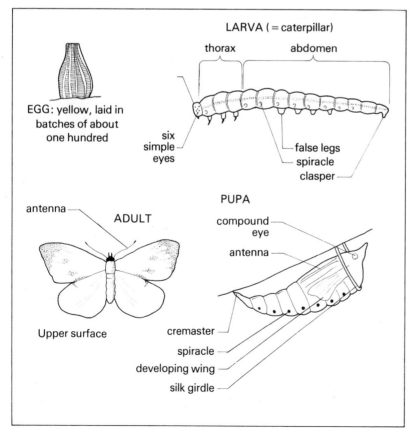

surface through its single breathing tube (spiracle). The larva moults three times. At the fourth moult the pupa emerges. The life history up to this point may take a few days or several months according to the temperature.

Unlike the housefly pupa, the gnat pupa is mobile. It does not feed but breathes from the surface through a spiracle on its head. When a few days old, it moves to the surface and the adult appears. Adult males feed entirely on nectar but the female gnat has to have a meal of blood before laying her eggs. Gnats are members of the family of mosquitoes. The female habit of bloodsucking plays a vital part in the transmission of tropical diseases such as malaria and yellow fever.

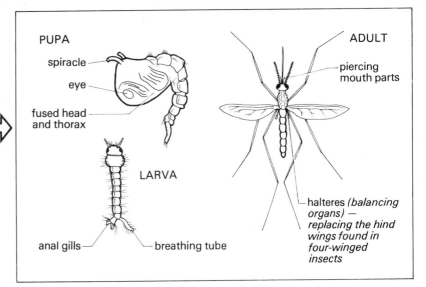

The dragonfly
Adults are seen in summer near ponds and streams. They feed on the wing, their diet consisting mainly of gnats. The young insects are completely aquatic. They look very different from the adult. They are called nymphs but unlike the butterfly they have no pupa stage. The eggs are laid on the water surface or within the tissue of a water plant. They take from four to six weeks to hatch or they may not hatch until the following spring. Development may take as long as three years. The nymph may moult as many as twelve times before it is fully grown. Wings begin to appear at the sixth moult.

The nymph is a fierce carnivore and will attack anything that moves, even animals bigger than itself. It breathes through gills which line the hinder end of its food canal. These are ventilated by pumping water into and out of the back of its food canal. The nymph finally climbs a water plant into the air and the last moult releases the adult.

Dragonfly and dragonfly nymph in water (inset)

Insect and man
Insects are man's most serious rivals for food and materials. This is largely his own doing. In clearing a piece of land to grow broad beans, for example, man creates an ideal situation for the blackfly (bean aphid) which also likes broad beans. There are no weeds in nature, neither are there insect pests. Weeds and insect pests are plants and animals, growing and living where man doesn't want them. In both cases he has almost certainly made it possible for them to be there. Nevertheless, some insects do have certain value to man; for example:

As sources of materials Honey and beeswax are obtained from the honeybee. Silk is obtained from the caterpillar of the silkworm. Lac forms the basis of shellac, used in paint, and cochineal is used to colour the icing on cakes. Both of these are obtained from scale insects.

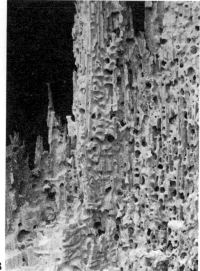

As pollinators Bees, wasps, butterflies and moths, together with some two-winged flies, are essential for the pollination of many flowers.

As predators on other insects Many insects actively feed on other insects. The ladybird feeds on plant lice (aphids). Dragonflies hunt other insects.

As scavengers Many insects play an important part in the breakdown of dead material. The maggots of houseflies develop in rotting materials. Dung beetles are important in pastures where cattle graze.

The chief disadvantages of insects from man's point of view are:

As general nuisances Flies in summer, ants, cockroaches, silverfish.

As destroyers of food and crops Grain (locust), cotton (cotton boll weevil), apples (codling moth).

As destroyers of materials Timber (termite, deathwatch beetle, furniture beetle), wool (clothes moth), books (book louse).

As carriers of disease Mosquito (malaria, yellow fever), tsetse fly (sleeping sickness), flea (plague). These insects transmit the disease when they bite or pierce the skin. The housefly is believed to be responsible for spreading contact diseases (cholera, summer diarrhoea and typhoid).

Bee visiting flower in search of pollen (A),
Timber damaged by death watch beetle (B),
Dung beetle – the dung is buried and eggs laid in it (C),
Burying beetles feeding on carrion (D)

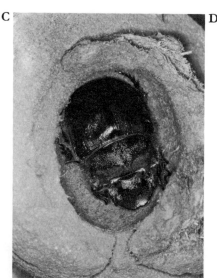

Principles of Biology

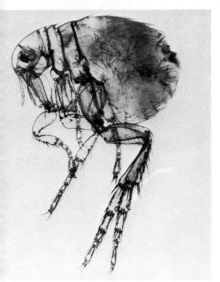

As biters or stingers Bees, wasps, ants, bedbugs, fleas, mosquitoes.

Flea – note the size of its jumping legs (E)

Mosquito taking blood meal through skin (F)

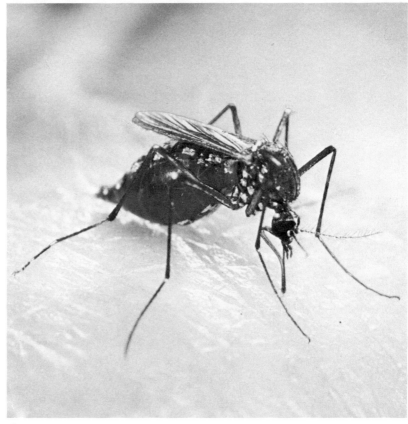

F

Arthropods

Insects belong to a large group of animals called arthropods. All arthropods have a dead outer skin which serves as a supporting exoskeleton. All possess jointed limbs. In addition to insects, the group includes the crustacea, the arachnida and the myriapods.

Crustacea (crabs, prawns, shrimps, lobsters, water fleas, woodlice) With a few exceptions, Crustacea are aquatic. They breathe by gills which are outgrowths of the body wall with a large surface area. The oxygen is circulated by means of a blood system. Crustacea have two pairs of feelers and many pairs of jointed structures, some of which are used as legs while others are used for dealing with food. The biggest crustacea are the lobsters and crabs. A full-grown lobster may be over fifty cm long and weigh several kilograms. It has a massive skeleton strengthened with calcium salts. Most of the group, however, are very small.

The surface layers of the oceans teem with small crustaceans. One group, called copepods, are believed by some people to exceed, in sheer numbers, all the other animals put together. Some crustacea are parasites while others, such as barnacles, spend their adult lives in one place. The woodlouse is able to survive on land by frequenting dark, damp places.

Arachnida (spiders, scorpions, mites, ticks)
These are land arthropods. The body is divided into two main

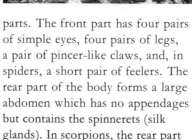

parts. The front part has four pairs of simple eyes, four pairs of legs, a pair of pincer-like claws, and, in spiders, a short pair of feelers. The rear part of the body forms a large abdomen which has no appendages but contains the spinnerets (silk glands). In scorpions, the rear part

Lobster (A),
Marine copepod (B),
Water flea (C),
Acorn barnacle feeding under water (D),
Woodlice on the underside of a piece of bark (E)

Principles of Biology

forms a jointed tail with a poison spine at the tip. Spiders feed mainly on insects which they catch in their webs. The prey may be paralysed with poison by some spiders. Others tie up their prey with silk. The spider injects a fluid into its prey. This digests its body contents, which the spider then sucks into its stomach. Almost all spiders are harmless to man and perform an important task in keeping down the numbers of insects.

Myriapods (centipedes, millipedes) Myriapod means many legs. These animals have bodies divided into many segments. Each segment carries jointed legs, one pair per segment in centipedes, two pairs per segment in millipedes. Centipedes are found in moist dark places, under stones, behind decaying bark and in the soil. They are carnivorous: they eat insects, earthworms and slugs. Millipedes are entirely vegetarian, feeding mainly on the dead remains of leaves in the soil. Despite their greater number of legs they move more slowly than centipedes.

Garden spider (A),
Scorpions (B),
Centipede (C),
Millipedes (D)

Molluscs

Garden snail (A),
Live mussel opened under water to show gills (B)

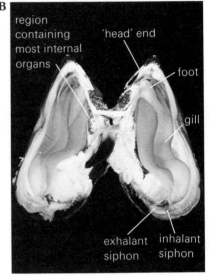

To the man in the street, molluscs and crustaceans are both 'shellfish'. However, molluscs are distinct in many ways; for example, their bodies are not divided into segments. Instead, every mollusc has three main parts – head, foot and body mass – all covered by a skin-like cloak called a mantle. Most molluscs have a limy outer shell. They are found in a wide variety of situations, but most live in or by the sea, especially on the shore or in shallow water. Some molluscs can float or swim but most move slowly by crawling; many attach themselves to other objects. A few burrow, even into rock.

Molluscs are important to man mainly as food (clams, mussels, oysters, etc.). Oysters provide pearls. Some molluscs damage marine structures such as piers. Certain tribes use mollusc shells as 'money'.

The animals described here belong to the three main groups of molluscs.

Snails and slugs (Gastropods)

These molluscs have a single, usually spiral shell which houses the body mass (slugs have a much smaller shell, placed just *under* the skin). The head is clearly visible when the animal emerges from its shell; the foot is larger than in other molluscs. The two pairs of tentacles on the head can be turned inside-out for protection.

Snails and slugs eat vegetation and may do much damage to cultivated plants. The tongue is covered with horny teeth, and works like a file against a bar in the upper jaw to rasp away at leaves. It grows continually as it becomes worn.

Each animal is both male and female. Mating is followed by cross-fertilization: the process is stimulated by a chalky dart which each animal shoots into the body of its partner. Sperm is transferred and eggs are laid in damp places; they are covered by a tough protective jelly. Small snails hatch from the eggs.

The soft body surface is protected by slime (mucus), which is also used in movement. Waves of muscle contraction pass along the underside of the foot; the animal glides over the mucus, leaving its characteristic trail.

Snails and slugs are most active at night and in damp weather. During the day they hide in their shells under rocks or in vegetation. Snails seal the shell entrance with slime.

Other close relatives of snails are the periwinkles and limpets of the seashore. All the water-dwellers breathe using gills, whereas land snails use a lung; this is situated in an air-filled space inside the shell.

Bivalves

These molluscs have thin bodies, as if they have been squeezed from the sides. The shell is in two parts (bivalve), hinged at the top. It completely encloses the soft body. The head is not distinct. The foot is wedge-shaped and smaller than in snails. The gills are extremely large, forming flat folds down each side of the body under the shell.

These features can be seen in the **freshwater mussel**. The shells are brown-green and show lines of seasonal growth. They are held together at the hinge by muscles and

ligaments. Water enters and leaves the animal through two siphons which can be pushed out through the partly-opened shells. The water passes over the gills, which are more important for feeding than for respiration. The gills have cilia which beat to collect tiny food particles in the water and pass them to the mouth.

Freshwater mussels can move slowly over the mud at the bottom of lakes, ponds, and rivers, by pushing out the foot and dragging themselves along. Marine mussels can anchor themselves to rocks, using threads which they secrete.

Most bivalves shed their reproductive cells into the water; after fertilization, there is a free-swimming larval stage. But in freshwater mussels, the females have their eggs fertilized inside their shells when sperm from a male enters with the water current. Eventually, the larvae are released to become temporary parasites on fishes, before becoming adult.

Squids, cuttlefish and octopuses
These marine molluscs have prominent heads surrounded by long mobile tentacles. The rather small shell is *inside* the body.

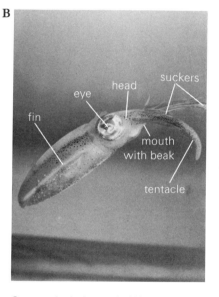

Compact bed of mussels (A), Squid (B)

Squids can be found swimming freely in coastal waters. The mouth is at the centre of ten tentacles – octopuses have eight – and has two horny jaws forming a powerful beak. The head is connected by a short neck to the slender tapering body. Below the neck, a muscular siphon lets water into a cavity under the body. Oxygen for respiration is extracted from this water. When water is forced out of the siphon, it provides jet propulsion. The siphon can be pointed in different directions to give different directions of movement. The tapered end of the body has a horizontal fin on each side, to assist swimming.

The skin contains pigment cells which alter the animal's colour; the colour is said to change with its emotional state. When molested, the animals are able to escape under cover of a jet of brown ink (sepia) ejected from the siphon. The flat shell (cuttlebone) is found just under the skin on the upper side of the body. It is sometimes given to caged birds for sharpening and cleaning their beaks.

Echinoderms (animals with spiny skins)

Echinoderms live only in or by the sea, on the shore (in rock pools, under seaweed or buried in sand), or near the surface of open waters, but especially on the ocean bed, even down to 3 000 fathoms. They have radial bodies – the body is divided into equal 'mirror' halves when bisected through the middle. Echinoderms have arms which radiate from a disc at the centre. The skin has a pattern of interlocking plates just under the surface, from which spines project outwards. The body is fairly simple. There are no segments. The nervous system has no brain. Echinoderms move slowly. Most reproduce by shedding eggs or sperm into the sea. After external fertilization, the larva which hatches from the egg swims freely in the plankton by means of its cilia, eventually settles and becomes adult.

The echinoderms most likely to be seen are starfish and sea urchins, but the group also includes brittle stars, heart urchins, sea cucumbers and sea lilies. Echinoderms are rarely directly important to man, although in some places they are used as food. Starfish sometimes cause great damage in commercial beds of clams and oysters; more recently, starfish have caused serious destruction to coral reefs.

Sea urchins

Sea urchins have rounded bodies and seem to have lost their arms. Actually, they are similar to starfish: a starfish could be 'turned into' a sea urchin by lifting its arms by the ends, gathering them up into a ball, and joining them together. The hard plates interlock just beneath the skin surface to form a shell, with interlocking movable spines standing out. There are small pincers between the spines.

The mouth and anus are placed at the centre – the mouth, with five teeth, on the lower surface, the anus on the upper. Openings from the reproductive organs are arranged in a ring around the anus. Similarly, gills for respiration are placed around the mouth. The mouth is supported by a hard framework called Aristotle's lantern.

Five double rows of tube feet extend around the sides from mouth to anus. They can be expanded and contracted by water pressure through holes in the shell. These five rows are equivalent to the five arms of a starfish.

Urchins and starfish live in similar situations. Some can bury themselves in sand and mud or soft rock. On the seashore, they hide at low tide in rock pools or under weeds. They move slowly using the tube feet (with suction pads at the ends) and jointed spines. They are scavengers, feeding on dead remains, but also on seaweeds and and small animals. Damaged urchins can grow new parts – shells, spines, tube feet, etc.

Starfish

Starfish can be found on most coasts; they are frequently stranded on sandy shores after storms. They can also be dredged from sand and mud at much greater depths. Colours commonly range through shades of orange and brown.

Five (usually) tapering arms radiate from around the central disc. New arms can be grown if

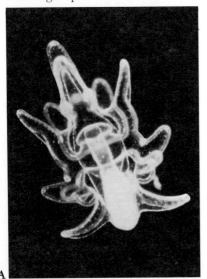

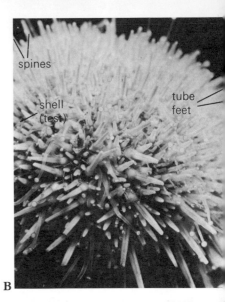

Ciliated echinoderm larva of starfish (A),
Upper surface of a sea urchin (B),
Undersurface of a sea urchin (C)

damaged. The upper surface of the rather flat body bears many blunt limy spines. Between these spines there are thin-walled gills for respiration and excretion, and also small, movable pincers. The pincers mainly protect and clean the body surface but may sometimes be used to catch food. The anus is near the centre of the disc.

The mouth is at the centre on the underside. A wide groove bordered by large spines extends down each arm and contains rows of tube feet used in walking and gripping. Each tube foot is a cylindrical sac ending in a suction pad; it can be extended or withdrawn by muscle action and by using fluid pumped in or out of the sac from the vascular system.

In order to move, the arms are raised, the tube feet extended to grip, and when they contract the body is pulled slowly forwards. If the animal is overturned, the arms are twisted so that some of the tube feet can get a grip to pull it upright.

Most starfish feed on other invertebrates, especially molluscs and crustaceans, though those living in deep water take in mud and feed on the organic matter which it contains. The typical method of feeding on a bivalve mollusc is for the tube feet to grip the shell and gradually pull it open. The stomach is then pushed out through the mouth to envelop the prey. Tentacles at the ends of the arms are sensitive to touch. Finally, the stomach with its partly digested food is taken back into the body.

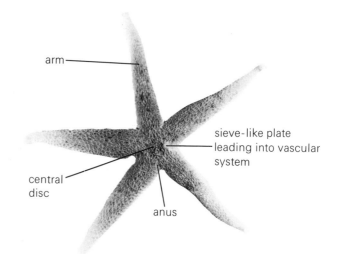

Upper surface of starfish

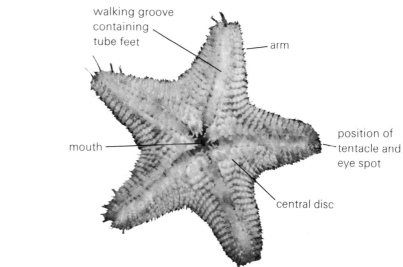

Undersurface of starfish

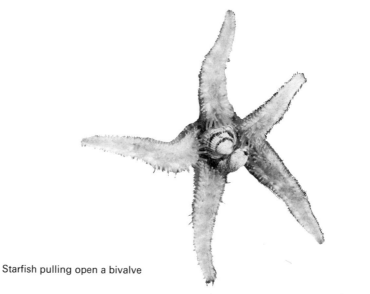

Starfish pulling open a bivalve

Echinoderms (animals with spiny skins)

Vertebrates

Vertebrates are animals with backbones. Biologists regard them as an important subdivision of a larger group of animals called the chordates. Chordates have, for at least part of their lives, a structure called a notochord which acts as a simple supporting rod; they also possess gills at some stage in their life histories.

In vertebrates, the notochord becomes a backbone. In land vertebrates, the gill slits disappear very early in life, except for one slit which becomes the tube joining the middle ear to the throat.

The main features of vertebrates are:

- An internal, living, bony skeleton, part of which is the backbone.
- Gill slits, leading from the throat to the outside, present for at least part of the life history.
- A highly efficient blood system, circulating blood within closed tubes. The blood contains an oxygen-carrying substance (haemoglobin) within special blood cells.
- A very complex nervous system and specialised sense organs including the eye and the ear.

Although the vertebrates are usually thought to consist of five classes (fish, amphibians, reptiles, birds and mammals), biologists divide them rather differently. They are split first into two groups – vertebrates without jaws and

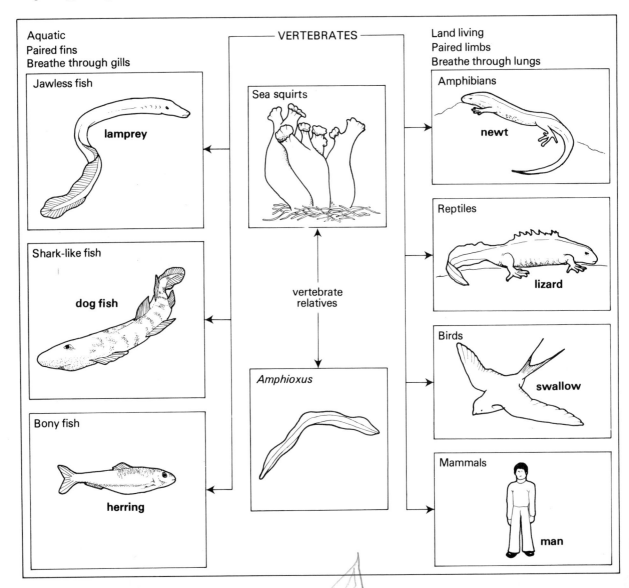

32 Principles of Biology

vertebrates with jaws and teeth. Those without jaws are a very small group. Some live in fresh water. These are the lampreys. Those living in the sea are called hagfish. Eel-like, they have many gills and prey upon fish to which they become attached by a sucker-like mouth, rasping away the flesh with horny rough tongues.

Vertebrates with jaws form two major groups. Those which live on land and possess limbs form the four classes of Amphibia, Reptiles, Birds and Mammals. Those which are aquatic and possess fins form three classes:
1. Sharks, skates and rays: fish with rough skins, separate gill openings and skeletons made of cartilage.
2. Bony fish; these include most of the familiar fish.
3. Lungfish: fish with swim bladders connected to the throat and which can be used to breathe air.

Each of these groups of fish is as different from the others as is the case with the land vertebrates.

Origins of vertebrates

The first vertebrates lived about 400 million years ago. Small, fish-like animals, covered with tough bony plates lying just below the skin, they lived in river estuaries. The backbone, providing a base for attachment of muscles, formed the basis for efficient movement, preventing them from being swept out to sea. The bony armour may have helped to protect them against the changing salt content in estuarine waters as well as their large invertebrate neighbours and predators.

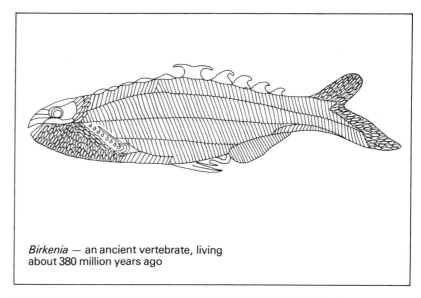

Birkenia — an ancient vertebrate, living about 380 million years ago

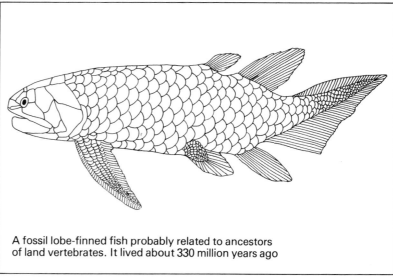

A fossil lobe-finned fish probably related to ancestors of land vertebrates. It lived about 330 million years ago

Vertebrates 33

Fish

There are two main groups of fish. The *bony fish* are found both in the sea and in fresh water. Their gills are protected externally by a plate called the operculum and their bodies are covered by thin overlapping bony scales. They possess swim bladders. They hunt their food, usually other animals, by sight and their eyes are found close to the mouth which is right at the end of the body.

Cartilaginous fish (sharks, skates, rays) are found only in the sea. They have skeletons made of cartilage (gristle) and the skin is covered by tiny scales which have the same structure as teeth. There is no operculum, the gills opening separately at the surface. They have no swim bladders and hunt chiefly by smell. The mouth is situated on the underside of the head.

Most bony fish shed eggs and sperm into the water where fertilization takes place. The cartilaginous fish lay few large eggs which are fertilized internally. They are protected by a horny case.

Life in water

Water is dense and almost incompressible. It offers support but also resistance to movement. Fish skeletons form a small part of the body weight since they do not have to support the body. The streamlined shape offers little resistance to the water and the hard, smooth surface reduces friction to a minimum.

A fish needs oxygen. Water contains little oxygen, and a continuous flow is required to provide

Shark showing mouth and gill openings (A),
Mermaid's purse — the egg case of a dogfish anchored to seaweed (B)

it in sufficient quantities. Water passes through the gills, which are slits in the body wall leading from the throat. The gills are lined with delicate filaments with a large surface area. Blood flowing through the capillaries of the gills is therefore exposed to a continuous flow of oxygen-carrying water.

Water contains salts in solution. The body fluids of a fish also contain a solution of salts, and its gills act as semipermeable membranes.

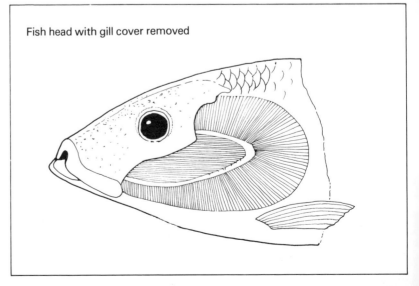

Fish head with gill cover removed

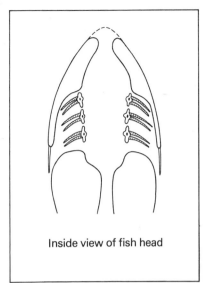

Inside view of fish head

In the concentrated salt solution of the sea, the fish loses water by osmosis. It therefore drinks large amounts of seawater, eliminating the salt through glands in its gills. In fresh water, the fish gains water and gets rid of it through its kidney.

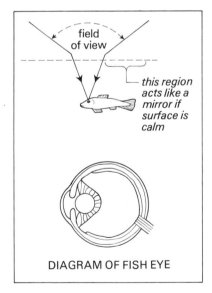

DIAGRAM OF FISH EYE

An eye immersed in water depends entirely on its lens for converging the light rays to form images. A fish lens is completely spherical and focusing is brought about by movement of the lens backwards and forwards in the eye.

Brown trout

The trout

The brown trout is found in lakes and streams all over the British Isles. Its size depends upon its age and its colour may vary with the composition of the water. It thrives in water containing plenty of dissolved oxygen. It is carnivorous and will eat anything that moves. Its diet includes water insects, snails, other fish, tadpoles and newts. Its body is covered with thin bony scales and the rings on these are a measure of age.

Contraction of the body muscles, first on one side and then on the other, make the tail move from side to side, driving the fish forward. Rolling is prevented by the dorsal and ventral fins; the paired fins are used to control movement up or down in the water. A swim bladder, containing oxygen, is found in the body cavity, just below the backbone. By secreting oxygen from the blood into the bladder, or removing it, the fish regulates its own density so that it remains at any depth without effort.

The trout breathes by means of gills. The raising or lowering of the floor of the mouth acts as a force pump. As it is lowered, with the mouth open, the reduced pressure causes the operculum to shut and water to enter through the mouth. As it is raised, with the mouth shut, the raised pressure drives the water out through the gills, pushing open the operculum. Oxygen diffuses out of the water into the gills, and carbon dioxide passes into the water from the gills.

Fresh water has a much lower concentration of salts than the body fluids of the trout. Some water therefore enters the gills by osmosis. The trout gets rid of this by producing large quantities of dilute urine from its kidneys. Ammonia, formed from the breakdown of protein, diffuses out of the body across the gill surface.

Trout, like most bony fish, hunt by sight. They have a poor sense of smell. The lateral line marks the position of a tube, containing sensory cells, just below the skin surface. It is sensitive to pressure changes. The internal ear is sensitive only to balance and movement.

Trout begin to breed after two years. They move upstream from the lakes and larger rivers. Spawning occurs in winter. Males and females pair off and, after an elaborate courtship, the female lays her eggs in a pit in the gravel bottom while the male sheds his sperm on them. About three months later the young trout hatch with the remains of the food supply attached as a yolk sac. After two or three weeks this finally disappears.

Young trout with egg sac just after hatching

Fish 35

Adaptation in fish

Although fast-swimming fish are all much the same shape, fish as a whole come in all shapes and sizes. They may be almost round (globe fish), or long and thin (pipe fish or eel). They may be flattened from side to side (plaice or sole) or from top to bottom (skate). Some fish lay few eggs and guard both the eggs and the young that hatch from them (stickleback), or they may lay millions of eggs which are left to take their chance in the surface waters of the ocean (cod). Many fish migrate, either to lay their eggs on the bottom in shallow waters (herring), or as part of their life cycle (salmon, eel).

A single scale from a three-year-old trout

Globe fish

Bottom dwellers are able to change colour to match their surroundings. Skates and rays cannot see food entering their mouths which are on the under-surface. Plaice and sole start life with an eye on either side of the head. One eye moves over to join its fellow.

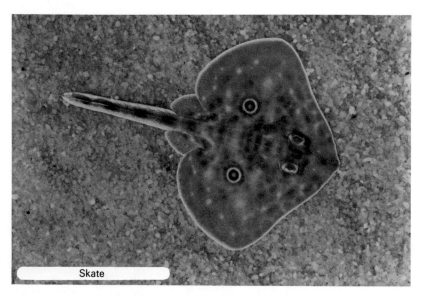

Skate

Sole

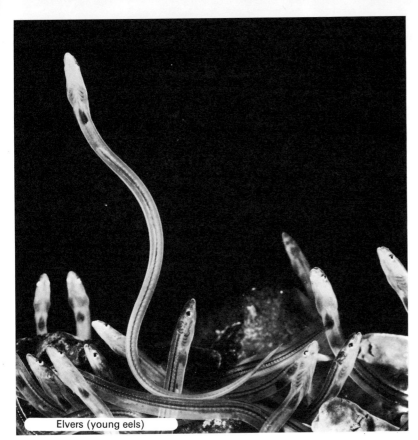

Elvers (young eels)

Eels are found in the rivers of the British Isles but only as adults. They swim downstream, into the sea. They reach their breeding grounds to the south of Bermuda. After laying their eggs, they die. The young eels, hatching from the eggs, set off and three years later reach the freshwater streams of Britain. After some years they in their turn set out for the breeding grounds.

The African lungfish lives in freshwater. Its swim bladder connects with the throat and can be used as a lung.

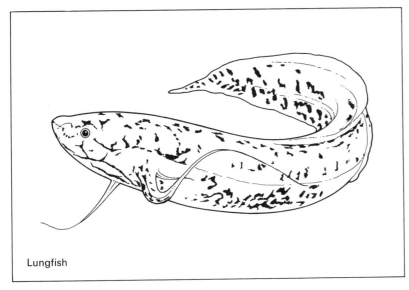

Lungfish

Amphibians

These are vertebrate animals, most of which can live on land and in water. They all depend on water for reproduction. Amphibians were the first land vertebrates and provide a link between fishes and reptiles. They have paired limbs instead of paired fins. They possess lungs, and their sense organs are developed for life both on land and in water. The skin is kept moist by the secretions of mucus glands – there are no scales. The body temperature changes with that of the surroundings; that is, they are 'cold-blooded'. Amphibians lay eggs covered by jelly but no shell; they hatch into larvae, usually water-living and breathing by gills. The larvae then undergo distinct changes, called metamorphosis, before becoming adult.

Newts and salamanders
These amphibians have bodies with a quite distinct head, trunk and tail, and their limbs are about equal in size.

Newts have rather slender bodies. They live, like frogs, in damp places for most of the year and go to the water in spring to breed. The male develops a prominent dorsal crest and vivid body colours to attract the female. The eggs are fertilized internally and then laid with great care singly, two or three times a day for several weeks. The female lays her egg on the leaf of a water plant which she then folds. The sticky jelly covering prevents the leaf unfolding. Newt tadpoles hatch from the eggs and resemble frog tadpoles except that the fore limbs develop before the hind limbs. Also, the newt tadpole keeps its external gills until it leaves the water.

Newts are carnivorous. They feed on worms, insects and molluscs. They can be mistaken for lizards, but they have moist skins without scales and the limbs do not end in claws.

Salamanders live in rivers, marshes and moist woodlands, concealed beneath stones or logs. A few species remain in water throughout their lives and keep their larval gills: their lungs never

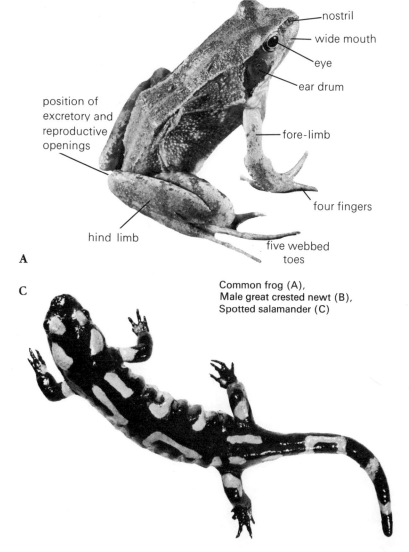

Common frog (A),
Male great crested newt (B),
Spotted salamander (C)

develop. There is no native British species though salamanders are quite common in temperate areas of Europe, North America, etc. The axolotl of Central America is a salamander which is a permanent gill-breather. Normally it remains at this larval stage yet becomes sexually mature and able to reproduce itself.

Frogs and toads

The bodies of frogs and toads are streamlined and rather triangular: there is no neck or tail. The fore limbs are much shorter than the hind limbs. The ear drums are visible on the head.

Frogs live in water or moist places nearby. They hibernate in mud at the bottom of ponds and in thick, damp vegetation. They emerge in spring when temperature and day-length increase, and return to water to breed. Male frogs call the females, and male and female swim around together, the male clasping the female by means of a thick horny pad which develops on each thumb at this time. The male helps to squeeze the eggs out of the female and then sheds sperm into the water to fertilize the eggs externally. A secretion from the female's oviduct covers each egg. On contact with water this covering swells to form a protective jelly, bringing the spawn to the water-surface where the eggs obtain oxygen from the air and warmth from the sun. The developing tadpoles inside the eggs feed on yolk which continues to feed them for several days after they wriggle free from the spawn. The tadpoles (which hatch after about a fortnight) are poorly developed, with the mouth and eyes not yet opened.

Methods of feeding and breathing change during the tadpole's development.

Feeding When the mouth has opened, a few days after hatching, the tadpole feeds on water weeds. It settles down, using its sucker, and rasps food off stones with its powerful horny jaws. The long coiled intestine can be seen through the skin of the abdomen. When the tail has grown longer and stronger the tadpole is able to chase and eat other water animals, even other tadpoles.

Breathing The young larva breathes through its skin and through tufts of external gills at the sides of its head. These gills disappear later and the tadpole breathes like a fish, taking water in through the mouth, passing it over internal gills, and out through a spout called the spiracle. Lungs are almost the last organs to develop and the tadpole then visits the surface more frequently, to breathe.

The long hind limbs develop from buds near the anus. Slightly later, the fore limbs grow out from the gill region. Finally, the tail gradually disappears, eaten from within by white corpuscles. The larva moves to the land and casts its skin.

Adult frogs catch worms and insects with their sticky tongues which are hinged at the front end. Often, insects are caught in flight. Whilst swimming and during hibernation the skin is used as the main breathing organ. On the land the frog uses the moist lining of the mouth cavity; during more violent movements, air is taken into the lungs. The webbing between the toes provides a large flat surface to push against the water during swimming. The length of the hind limbs makes possible a powerful thrust against the ground when leaping.

Newly hatched tadpoles (A),
Tadpole with internal gills (B),
Late tadpole with four legs (C)

A
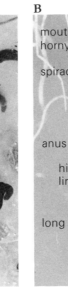

B
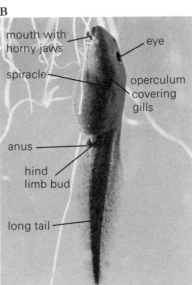

C

Reptiles

Present-day reptiles are all that remain of what was (about 200 million years ago) a very large, dominant group of land vertebrates. Now there are only three main types of reptile – snakes and lizards, turtles and tortoises, and crocodiles. Reptiles, like amphibians are cold-blooded – they are particularly abundant in tropical and subtropical regions, and most reptiles need quite high temperatures before they become active. Unlike amphibians, they are reasonably well adapted to life on land. The skin is dry, with protective scales. Reptiles breathe air, using lungs only. Fertilization is internal. When the eggs are laid they are protected by tough, horny shells. The eggs are fully provisioned with yolk to feed the developing reptile, so that it emerges at a late stage – there is no larval stage. Each egg has membranes enclosing a miniature 'pond', which helps to protect the developing reptile from drying and mechanical damage.

Crocodiles and alligators
These are tropical reptiles, found in America, Africa, India, North Australia and South-East Asia. They live in rivers and marshes, partly submerged for long periods. They are large animals, as much as seven metres from snout to tail. All parts – head, trunk and tail – are long. The tough leathery skin forms horny, closely-fitting scales over most of the surface. The jaws open widely to grip quite large prey, and are armed with many powerful pointed teeth. The throat can be closed when the mouth is open, so that water does not flood into the lungs. The toes are webbed. The tail is heavy.

Female alligators lay about fifty eggs at a time in nests built of vegetation. The eggs are warmed as the vegetation decays. Young alligators emerge from the eggs after about two months. The adults sometimes dig shelters in the banks of rivers.

Turtles and tortoises
These reptiles have their bodies enclosed in a 'shell' composed of closely-fitting plates covered by hard scales. The shell is in two parts, one above the body, the other below. The plates of the upper part are firmly attached to the ribs and backbone. The lower part is much flatter. The two parts are widely opened at the front (for the head and fore limbs) and at the back (for the tail and hind limbs). All these softer parts of the body can be withdrawn into the protective shell. There are no teeth but the horny jaws can crush food. The limbs end in claws.

Land-living tortoises have short, stumpy limbs spread out at the

Crocodile

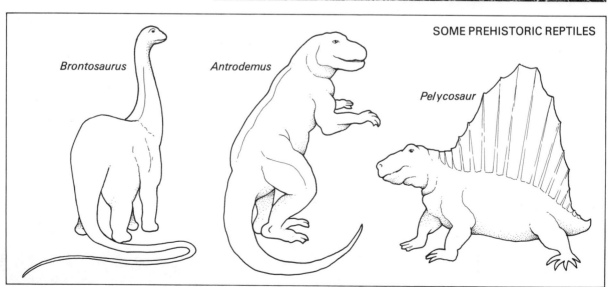
SOME PREHISTORIC REPTILES
Brontosaurus
Antrodemus
Pelycosaur

Principles of Biology

Tortoise (A),
Chameleon (B),
A lizard shedding its skin (C),
A grass snake showing its forked tongue (D)

sides of the body, making movement slow and awkward. Turtles are water-dwellers; their limbs form flippers with webbed toes. They live in streams, marshes and in the sea. They breathe with gill-like structures at the end of the digestive system. Marine turtles come on land only to lay their eggs. Freshwater turtles hibernate in the mud at the bottom of ponds and lakes.

Lizards and snakes
These reptiles are found in almost every habitat, in a wide variety of shapes and sizes – very large monitor lizards, found in Asia, may reach four metres long. The skin is covered by many small scales, is moulted at intervals, and is capable of stretching when large prey is swallowed. In most lizards, the skin is loose and flexible.

Some lizards have small limbs, others such as slow-worms (found in Britain) have no limbs at all. These are mainly soil-dwellers and move by making S-shaped movements of the whole body. Fast-moving lizards are counter-balanced by their long, slender tails, but many other lizards have short tails. The joints between the tail-bones are relatively weak. If a lizard is seized by the tail the bones can separate and the hind part of the tail is detached – the lizard may run to safety, leaving its wriggling tail.

In many lizards, the tongue is pushed out of the mouth to be used as an organ of taste and touch. In chameleons, the tongue has a sticky tip and can be extended very quickly to trap insects for food. Chameleons have other special features: each eye can be moved independently, and they have grasping hands, like those of man, but with two fingers working against three.

Snakes are limbless reptiles showing many special features. There are no outer openings into the ears. The mouth is armed with backward-slanting teeth, both on the jaws and in the roof, to hold the live prey before swallowing. Non-poisonous snakes – pythons, boas and the grass-snake – coil around their prey and kill by suffocation. The 'fangs' of poisonous snakes are a pair of special teeth in the upper jaw which inject venom to immobilise prey. The jaws are connected at the front by ligaments only, so that they are extremely elastic. Some snakes can swallow eggs twice their own body diameter. The 'fangs' of vipers and rattlesnakes are more efficient than those of cobras because they are longer, nearer the front of the jaw and can be folded back when not in use. The venom flows down a tube in the tooth. However the black-necked cobra can accurately spit venom and this may cause serious eye damage.

Snakes usually move by looping the body sideways and pushing these loops against uneven surfaces. They can also use their scales to push against the ground. The opening into the windpipe is much further forward than in other reptiles. They can swallow their prey whole and breathe during the long process of swallowing. Most vipers have a pit between nostril and eye. This is sensitive to temperature changes and is used to detect warm-blooded prey.

Lizards and snakes become dormant in cold weather, seeking shelter in burrows or crevices.

Birds

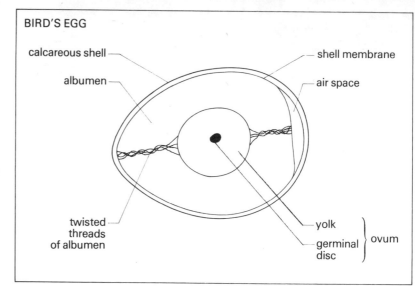

Like mammals, birds are descended from ancient reptile ancestors. They are mostly small and basically alike. Body structure and muscle needed for flight leave little room for variation in basic shape.

Birds have a high, constant body temperature (40°C) and the feathers are very efficient as insulators against heat loss. Upper and lower jaws form a beak, or bill, and there are no teeth. The lungs are extended into air sacs; when the lungs are emptied and filled, the air in the lungs is completely changed. The breastbone is much enlarged, to act as anchor for the massive flight muscles which move the wing. These muscles may account for up to half the total weight of the bird.

Most water loss occurs in air which the bird breathes out. The kidney excretes very little water; the almost solid uric acid produced from protein breakdown leaves the body, together with the waste from the digestive system, as droppings.

The egg is fertilized internally. When birds mate, the single rear opening (vent) of the male is applied to that of the female and semen containing sperm is passed into her body. After the sperm cell has fused with the egg cell, a layer of albumen (egg white) is added by the wall of the egg duct, and finally a shell.

The nest may be a complex affair, or a heap of twigs. Sometimes there is no nest at all. Concealed or inaccessible nests contain pale-coloured eggs. Others are either camouflaged or brightly coloured and therefore easily seen. The brightly coloured ones usually have a nasty taste, so the colour acts as a warning to predators. The yolk in the egg is the food store for the young chick. The shell allows the diffusion of gases so that the young bird is able to breathe. The eggs are kept warm (incubated) by one or both parents, which form mating pairs. Birds do not form groups of females dominated by a single male, as some mammals do.

Birds show complex forms of behaviour in their nest-building

Great tit's nest

42 **Principles of Biology**

Mother blackbird feeding her young

and courtship behaviour. Many birds migrate in the autumn to warmer or otherwise more favourable regions. The mechanisms by which they find their way are still largely unknown.

The blackbird
The blackbird is found almost everywhere, in woodlands, hedgerows, city parks and suburban gardens. The male is jet black and has a bright orange beak. His mate is a rusty brown colour. Both sexes are aggressive and frequent disputes occur between adult birds, especially just before the breeding season.

Blackbirds feed on earthworms and insects, caterpillars being a special favourite. Town birds will readily eat household scraps. The female begins to build her nest in February. She builds it in bushes or on ivy-covered walls, using twigs and grass. The nest is cup-shaped, the inner surface being lined with wet mud and a final layer of fine grass. Mating is preceded by an elaborate courtship.

The eggs, which are pale green with brown speckles, are laid from March onwards, between three and five at a time. Incubation takes about thirteen days. The task of keeping the eggs warm is made easier by the presence of featherless patches (brood patches) on the underside of the parent birds.

When the young hatch, both parents are kept busy finding food for their offspring. After about a fortnight the young are ready to leave the nest, but for some time afterwards they are still dependent upon the parent birds for food. Two or three broods may be raised between March and July.

The male blackbird has a beautiful song which is often heard at dusk as he stands on a roof or treetop. At other times the birds may make a loud, rapid 'chook, chook' sound. This warning noise often means that a cat is nearby.

Blackbirds are permanent residents; they do not migrate.

Flight
Birds do not 'swim' through the air as is commonly believed. A bird's wing, like that of an airplane, is curved so that the air passing over it has further to travel than the air passing under it. The faster-moving air above the wing is more spread out and therefore is at lower pressure than the

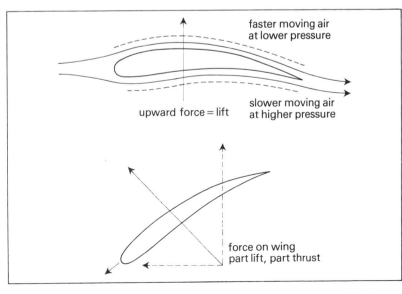

Birds 43

more compressed air underneath. It is this pressure difference that creates lift. The faster the air flow the greater the lift.

The inner part of the wing moves through a short distance only. It acts like the fixed wing of an aeroplane. The power stroke of the wing is *forwards* and downwards. The forward movement creates *lift*; the downward movement creates *thrust*. The recovery stroke is upwards and backwards; The main flight feathers twist and part to allow air to pass through the wing, thereby reducing resistance.

When a bird has a low airspeed, for example at take-off and landing, the power stroke, in order to create maximum lift, may carry the wings so far forward that they almost meet in front of the bird. A small structure on the front edge of the wing gives extra lift at low speeds, by projecting forward just as a 'slot' does on the leading edge of an airplane wing. When landing, a bird spreads its tail fan and holds it downwards. It acts like the flaps on an airplane, giving extra lift and also acting as a brake.

When a bird has sufficient speed through the air it is able to 'freewheel' or glide. It is able to maintain this speed by travelling 'downhill'. If the air through which the bird is travelling downhill is itself rising (under certain sorts of cloud, over hills, or at cliff faces) the bird may actually gain height while gliding.

Adaptation in birds

Although appearances may suggest otherwise, birds vary surprisingly little. This uniformity is imposed on birds because they are flying animals. The variation they do show is limited to their beaks, their feet and the wing shape.

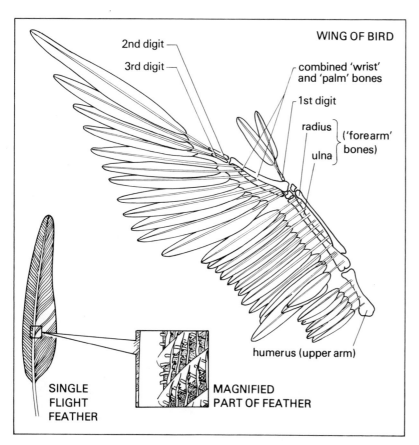

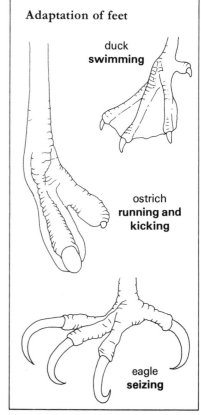

44 **Principles of Biology**

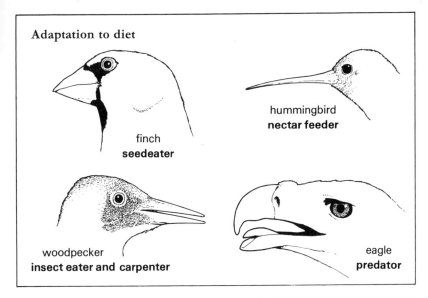

Flightless birds

Penguins When on land, penguins are clumsy and ungainly birds. In the sea, where they spend most of their lives when not breeding, they are superb swimmers. They are streamlined, using their wings as paddles and their feet as rudders.

Ostriches Although ostriches are found on the plains of Africa where large carnivores live, they are well able to take care of themselves by their large size and high running speed. They have a powerful kick. A full-grown bird may be 2.5 m tall and weigh more than 100 kg.

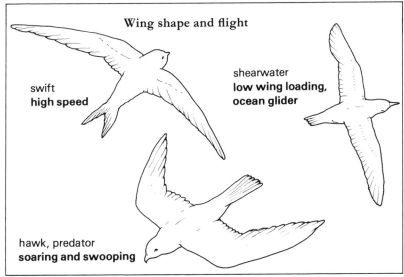

Birds 45

Mammals

Mammals are animals whose bodies are entirely or partly covered with hair. They suckle their young. A sheet of muscle (diaphragm) divides the body cavity into two parts. The diaphragm, together with the rib cage, forms the basis of an efficient bellows system for breathing. Mammals have external ears. Their large, complex brains enable them to show much more intelligent behaviour than any other animals.

Mammals are found all over the world: on land, in the seas and in freshwater; from the poles to the equator; in hot steamy rainforests and in the deserts. They are divided into three main groups:

Primitive mammals (monotremes) Only two species are known. They are the duck-billed platypus of eastern Australia, and the spiny anteater of eastern Australia and New Guinea. They are the only mammals that lay eggs; they suckle their young after they hatch. They have no teeth.

Pouched mammals (marsupials) These are found in Australia and New Guinea, in Central and South America, and one species as far north as the southern states of the

Kangaroo and koala bears (inset)

Spiny anteater

Duck-billed platypus

46 **Principles of Biology**

USA. They were much more widespread millions of years ago, but have been replaced everywhere else by later mammals. Their young are born at a very early stage of development. The young animal crawls into the pouch on its mother's abdomen. It attaches itself to a nipple and remains there until it has finished its development. Marsupials include the kangaroo, the koala bear and the Tasmanian wolf.

Placental mammals

Most mammals belong to this group, including humans. Like other mammals, the egg is internally fertilized. It develops inside the mother, the young creature being born at an advanced stage in development. Before birth, the developing young (foetus) receives food and oxygen from the mother through a special organ called the placenta.

Some of the features which have had a great deal to do with the success of the placental mammals are:

Their constant body temperature linked with a constant and very high rate of body activity (comparison with frog)

Their means of support and locomotion (comparison with lizard)

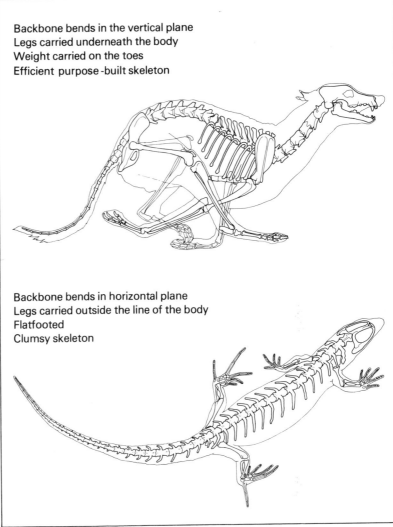

Backbone bends in the vertical plane
Legs carried underneath the body
Weight carried on the toes
Efficient purpose-built skeleton

Backbone bends in horizontal plane
Legs carried outside the line of the body
Flatfooted
Clumsy skeleton

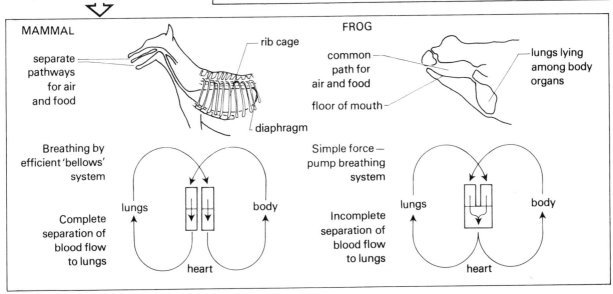

The protection and development of the young inside the mother's body (comparison with frog)

Protection by mother's body.
Young mammal develops at a constant rate because of constant temperature.
Constant food and oxygen supply through placenta from mother's blood.

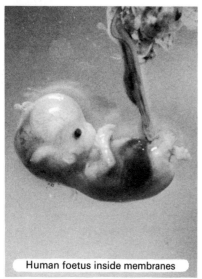
Human foetus inside membranes

Egg covered by jelly; slippery and hard to grasp.
Time of development variable because temperature of surrounding water varies.
Food limited to yolk in egg; oxygen by diffusion from water.

A complex skin, controlling both water loss and heat loss (comparison with frog)

Dead waterproof layer on outside.
Hair on the outside and fat in the dermis.
Sweat glands to eliminate excess body heat.

Skin section with hair follicles

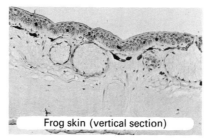
Frog skin (vertical section)

Living outer layer, non-waterproof.
Unprotected skin with no insulating material.
Mucus glands to slow down constant water loss.

Adaptation of mammals to different habitats

Fast running on open plains:
horse, zebra, deer
Leg length, and therefore stride, increased by running on the tips of the toes.
Eyes at the side of the head for

Zebra

all-round vision.

Soil life: mole
Dense thick fur to exclude soil particles.
Small eyes and close-set ears.
Short powerful spade-like limbs

Mole

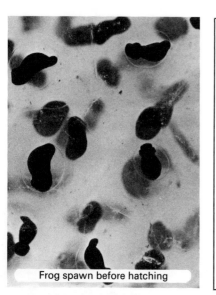
Frog spawn before hatching

Intelligence
Mammal's brain in comparison with that of a lizard

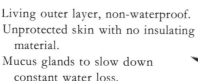 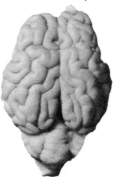

Low brain weight/body weight ratio
Small cerebrum

High brain weight/body weight ratio
Large cerebrum, co-ordinating action of the rest of the brain

48 **Principles of Biology**

Life in water: seal
Streamlined shape and reduced external ear reduces resistance to water.
Thick fat layer under skin to exclude cold.
Paddle-like limbs.

Adaptation to extreme conditions
Lack of water: desert rat
A diet of seeds containing oil; this produces more water as an end-product, when burnt in the body, than carbohydrate or protein.
Desert rats are nocturnal; they hide

High pressure and prolonged lack of oxygen: whale
Airway from blowhole (nostril) to lungs completely separate from food pathway.
Small lungs; most of the whale's oxygen, before a dive, is stored in its muscles and blood.

Grey seal

Whale

Gerbil (desert rat)

Living in trees: monkey (gibbon)
Eyes pointing in the same direction for distance judgement (stereo-vision).
Big toe and thumb separate from other digits for grasping.
Long grasping tail.

in the shade during the day, becoming active only at night.
Very little water is lost in the urine which has a higher salt concentration than sea water.

Extreme cold: polar bear
Large size; this means that the surface area, compared with its volume, is small and therefore heat loss, per unit of body weight, is also small.
Loose shaggy coat provides efficient insulation.
White colour helps prevent heat loss.
Its diet is rich in oils and fats (fish and seals).

Short, backward-pointing ribs, with diagonally placed diaphragm, makes the lung cavity long and narrow. Under great pressure, the lung cavity collapses without damage to surrounding structures.
In long dives, the blood bypasses the muscles but not the brain.
The muscles can work for long periods without oxygen.
Whales can tolerate high concentrations of carbon dioxide.

Gibbons

Polar bear

Flight: bat
Small light body.
Fingers long and thin to support a wing made of skin.
Large ears used in 'radar' system for detecting obstacles.

Bat in flight

Mammals 49

Bacteria

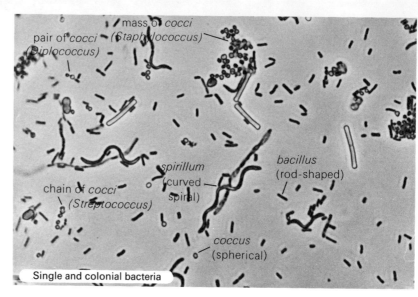

Single and colonial bacteria

Bacteria are single-celled microscopic organisms. They are the smallest known living organisms that have a cell structure. Even the largest are little more than $\frac{1}{100}$ mm long.

Bacteria exist singly or in groups of single cells. Bacterial cells have fairly simple shapes, which can be used to distinguish different types. Colonies of cells also have characteristic shapes. The photograph shows a collection of different bacteria as seen under the microscope. The smallest bacteria are *cocci*, less than one-tenth of the size of some *bacilli*. The diagram shows the features that may be present in a bacterium. There seem to be no large vacuoles, and in some bacteria the nucleus does not seem to be a single structure. Certain bacteria can make slow gliding movements; others can move using one or more flagella.

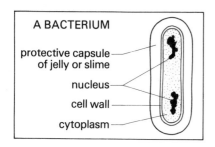

Bacteria are the most numerous living organisms. They are found in almost every situation (in the soil, in fresh and salt water, on and in other organisms) and under a variety of conditions (in the waters of hot springs or well below freezing). A handful of soil will contain many millions of bacteria. Being so small, they are easily carried by the air – bacteria have been collected several kilometres above the earth's surface. Others are carried by water. Some bacteria thrive where there is no oxygen.

In favourable conditions, bacteria

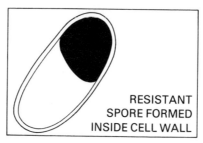

have an extremely fast rate of reproduction: one cell can divide at least once every hour. One bacterium could produce several millions each day. In unfavourable conditions, resistant spores may be formed inside the original cell wall.

Bacteria feed in three main ways:

1 **Free-living bacteria** Many are able to make their own food. This requires a source of energy. Some contain chlorophyll and feed by photosynthesis, using energy from sunlight. The nitrifying bacteria of the soil obtain energy from the chemical reactions which they cause.
2 **Saprophytic bacteria** These digest, using enzymes, the dead remains of living organisms or the substances that these organisms have produced.
3 **Parasitic bacteria** These obtain their food from the tissues of other living organisms: many cause serious diseases of man, his domestic animals and sometimes his crop plants. Examples of diseases caused by bacterial parasites are tuberculosis, anthrax, bubonic plague and pneumonia.

It is important to realise that although many bacteria are harmful to man – by causing disease, spoiling his food, etc. – many more have a beneficial effect in nature or are of direct use. There would be no fertile soils without the activity of certain bacteria. Bacteria are used in sewage disposal and in the decomposition of garden compost. Industrial processes (such as the production of acetic acid) and food production (such as the flavouring of cheeses) depend on bacteria. Some chemicals would be very difficult to make without the aid of bacteria. Many living organisms harmful to man are controlled by bacteria, some of which produce useful antibiotic substances.

Many bacteria live harmlessly in the bodies of animals, including man. Others are positively beneficial. Ruminant animals (those that chew the cud – cattle, sheep, etc.) would not survive without special bacteria which help to digest food that the animal cannot digest by itself.

Much of the direct harm caused by certain bacteria is due to their excretions. These may be poisonous (toxins) and responsible for the symptoms of disease (such as fever, inflammation) and for the effects of poisoning when foods become contaminated.

The work of Pasteur

Louis Pasteur, a professor of chemistry who lived a hundred years ago, was asked by the French government to investigate important diseases. People once believed that bacteria came from non-living particles, from dust in the air. Pasteur's experiments showed that bacteria, like all living organisms, came from the reproduction of other bacteria. His apparatus is shown in the diagram. The broth was made sterile by heating in the flask. The bacteria from the air could reach the broth only by passing down the tube; in fact, they were trapped at the bend. (A similar arrangement, called a 'fermentation lock', is used to keep out bacteria in wine-making.) No bacteria grew in the broth, even after a very long time.

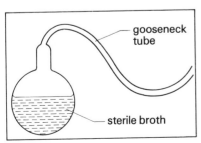

Many other scientists have made important advances in the study of bacteria, and have helped to fight diseases. Robert Koch devised methods for growing bacteria in cultures; Joseph Lister developed antiseptic surgery.

Food preservation

All living organisms require certain conditions to grow and reproduce. Most methods of food preservation involve removing one or more of these conditions. Bacteria, present in all foods, are then unable to spoil the food, as they are not present in large enough numbers.

1. **Refrigeration** If a low enough temperature is maintained, most bacteria will not reproduce. Food decay will be slow. Meat can be kept at around $-11°C$ but other foods must be stored at lower temperatures to preserve their nutritional value and vitamin content.

   Food rapidly decays when removed from refrigeration because sudden warmth stimulates the bacteria.

   Vacuum packing is a modern method of marketing food – air is withdrawn and the packet sealed before the food is refrigerated. Foods treated in this way have a very limited 'shelf-life'.

2. **Dehydration** Dried foods can be kept indefinitely if the storage place is not damp. Preservatives such as sulphur dioxide are sometimes added. A modern method, 'freeze-drying', combines the advantages of dehydration with refrigeration.

3. **Salting and jam making** Bacteria cannot survive in concentrated salt or sugar solutions because they lose water by osmosis. Beef can be transported, unrefrigerated, in brine, though its flavour alters slightly. Fruit can be preserved in sugar, but this produces food with a much higher calorific value. Drying fruit – grapes to form raisins, etc. – also makes the sugar content too great for bacterial survival.

4. **Pickling** Bacteria cannot survive in acid conditions. Acetic acid (in vinegar) provides a suitable acid preservative.

5. **Smoking** Herrings (to produce kippers) and ham can be cured by suspending them over smouldering wood chips. The food is partly dried by the heat. It is also preserved by the creosote materials in the smoke and by a layer of carbon deposited over the surface.

6. **Canning** Food is heated to high temperatures, often well above $100°C$; this kills active bacteria and their spores. The food is then packed in cans and the air expelled. The cans are reheated, immediately cooled, and finally sealed so that no more bacteria can enter.

7. **Bottling** When fruit is bottled, the temperature is raised only to $100°C$, but this and the acids from the fruit are enough to kill bacteria. Each bottle is covered by a metal disc made airtight with a rubber washer, and held on tightly. When the air in the neck of the bottle cools, it contracts and creates a vacuum. No more bacteria can enter.

   (Fruits with hard skins, such as apples and plums, can be preserved for several months if kept cool. The fruit should not be over-ripe but should be unbruised with intact skins.)

8. **Sterilization** Milk sterilized by boiling eventually turns sour because, although active bacteria are killed, their spores remain alive. Unfortunately, boiling changes the proteins in milk and

alters its taste. (In hospitals, all instruments are steam-sterilized under pressure.)

9. **Pasteurization** This process, originated by Louis Pasteur, kills all the most dangerous bacteria in milk without altering the flavour. The milk is either heated to 63°C for thirty seconds or to 72°C for fifteen seconds. Immediate cooling to 10°C slows the growth of the surviving bacteria.

10. **Radiation** This is a new method – food is packed in sealed containers and then bombarded by radioactive waves to kill the bacteria inside.

A

B

Haddock are placed to cool on trestles after removal from a sealed smoke barrel containing smouldering oak chips (A), In a canning factory waiting for the canned produce to emerge from the pressurised ovens (B)

Principles of Biology

Viruses

Viruses are the smallest and simplest of all known organisms. They are able to live only as parasites inside the living cells of other organisms. They cause some of the most serious diseases of animals – foot-and-mouth (caused by the first virus to be discovered), influenza, polio, measles, chickenpox. Viruses also cause diseases in crop plants – leaf roll in potato, mosaic disease of tomato and tobacco, spotted wilt in dahlia and tomato. Greenfly (aphids) are known to carry many viruses from plant to plant. In man and other animals, virus diseases are caught mainly by close contact with already-infected individuals. Infection may be through the air (by sneezing – common cold) or through contaminated food or food containers. Some plant viruses have only minor and apparently harmless effects, such as altering the colour of the leaves.

Outside the cells of their host, viruses seem to be no more than chemical molecules: an outer protein coat with another chemical (called a nucleic acid) inside. They have very simple shapes – spherical, brick-shaped, or long and thin. Special viruses that attack only bacteria have a 'head' and 'tail', rather like a short drumstick. Viruses become 'alive' and are able to reproduce inside other living cells. They do not divide themselves, but instruct the host cells to make more virus. The cells eventually die, setting free the newly-formed virus particles.

Typhoid virus (A),
Vaccinia virus (B)

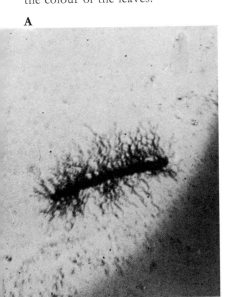

Algae

These are simple plants. Some resemble protozoans and are single-celled; others form long, fine threads. The largest algae (many of the seaweeds) consist of flat expanses of tissue. All algae feed themselves by photosynthesis. Many contain chlorophyll and are green. Other algae feed themselves using coloured substances which take the place of chlorophyll. Carbon dioxide and water are taken from the surroundings and sugars are made.

Single-celled algae
Chlamydomonas is common in ponds and ditches. It swims freely, by lashing its flagellum, to places where there is enough light for photosynthesis. The pyrenoid stores starch from photosynthesis in the chloroplast. *Asexual reproduction* takes place when the nucleus and then the rest of the cell contents divide into several spores; these grow flagella and swim away when the parent cell wall breaks open. In favourable summer conditions, one cell may give rise to two million offspring in a week, turning the water green. A similar process takes place during *sexual reproduction* except that the swimming bodies are gametes and have no cell wall. They fuse in pairs to form swimming zygotes each with four flagella; each eventually rests in a protective wall. Later, after being dispersed by wind or on the feet of wading birds, the zygospores divide, forming several new single-celled plants.

Pleurococcus is not found in water but in damp situations on land – walls, fences, tree trunks, especially on the north side. It can be seen as a powdery green covering. Apparently, it reproduces by asexual reproduction only, the cells dividing into two and then four. In favourable, very moist conditions, the offspring do not separate. In time, they form a colony.

Seaweeds
These, the largest algae, are more obviously like plants and even have root-like parts. There are three main groups:
1 **Green seaweeds** Found mainly between high and low tide levels, in rock pools and on sheltered sandy shores.
2 **Red seaweeds** Found on sheltered rocky shores and in deeper water.
3 **Brown seaweeds** Found on exposed rocky shores, but also beyond the low-tide level and freely floating in the ocean: some are the largest seaweeds, more than sixty metres long.

All three types contain chlorophyll, but the reds and browns have other pigments that help them to photosynthesise in deeper water where the light is dimmer and bluer than normal.

Seaweeds provide essential food for many marine animals and, in addition, those on rocky shores give shelter from rough seas and from drying when the tide is out. Many different marine creatures are to be found amongst seaweeds on rocks and in pools.

Seaweeds are now farmed as the basis of an expanding modern industry. They are used as fertilizers. The jelly-like materials (alginates) in seaweeds are important parts of a variety of foods from ice cream to salad cream, and are used in making adhesives, plastics and textiles.

Bladder wrack, a common brown seaweed, is firmly attached to rocks by its holdfast. Its stalk and thick midrib strongly resist wave action

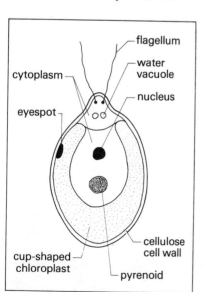

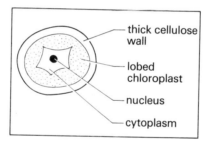

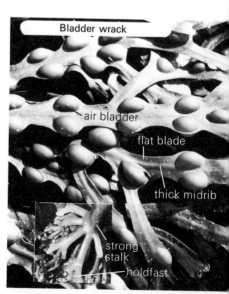

on the shore, and the mucilage that makes it slimy helps to retain water when the tide is out. The air bladders float the plant nearer to the light when submerged.

The swollen ends of the branches contain many openings; male and female reproductive organs line the flask-shaped cavities inside. When the tide goes out the plant dries and contracts, and packages of sperm and eggs are squeezed through the openings. When the tide next comes in, the packages are dissolved and the gametes set free. Each sperm has two cilia and can swim to the eggs. The eggs pro-

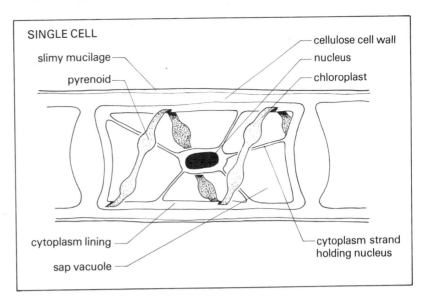

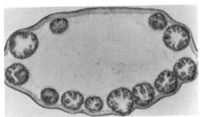

Bladder wrack in section showing cavities with sexual organs

duce chemicals to attract the sperm. The fertilized egg immediately settles, germinates and forms a new seaweed.

Filamentous algae
Spirogyra floats at the surface of ponds and slow-flowing streams. Its thread (filament) is composed of cylindrical cells joined end-to-end. It does not move except when the water is disturbed. In daylight, the chloroplast makes sugars which are changed to starch and stored in the pyrenoids. The cells grow longer until the nucleus and cell contents divide into two smaller cells, with a new cellulose partition in between. The whole filament grows longer. Another form of asexual reproduction, called *fragmentation*, occurs when splits in the cellulose partitions break up a filament into shorter lengths.

Sexual reproduction begins when two filaments lie side-by-side. The

Mucilage prevents organisms attaching.
Vacuole contains dissolved food substances.
Chloroplast is a flat green ribbon spiralling through cytoplasm lining.

cellulose wall grows out sideways where the cells of one filament touch those of the other. The outgrowths then link to form tubes connecting the cells of the two filaments. The contents of all the cells in one filament round off, forming gametes. Each gamete passes down its tube and fuses with the now rounded-off contents of the opposite cell. Each fused mass develops a thick protective wall and these zygospores are set free when the parent filaments break up. They fall to the bottom of the pond and germinate when conditions are favourable. Then the wall breaks, two small *Spirogyra* cells appear and float to the surface.

Diatoms
These algae contain chlorophyll but the green is masked by a brown pigment. The cell wall is hardened with silica (a material also found in glass, sand, and granite) and is in two parts, one overlapping the other. The surface is patterned.

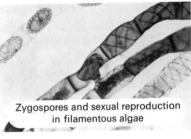

Zygospores and sexual reproduction in filamentous algae

The shape of the cell wall and its surface markings is different for each species of diatom.

Very large numbers of diatoms are found in plankton, where they provide food for many fish; diatoms are therefore important in food chains.

When diatoms die, their hard walls may form fossils. In some parts of the world their remains have formed fossil deposits of diatoms several hundred metres thick. These are mined and used industrially for insulation, filtering, and for scouring materials (for example, in toothpaste).

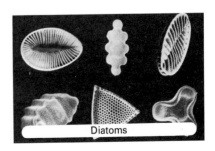

Diatoms

Algae 55

Fungi

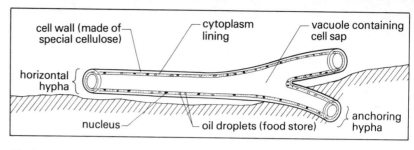

These simple plant-like organisms have no stem, leaves or roots. Instead, each fungus consists of a mass (a mycelium) of minute tubes called hyphae. Unlike most other plants, fungi do not contain chlorophyll: many, such as mushrooms, flourish in the dark. Some fungi are brightly coloured – orange, blue, green – but the green is not chlorophyll and fungi cannot make their own food by photosynthesis. All fungi use ready-made food which they obtain by one or both of the following methods:
1 By digesting the dead remains of organisms – that is, they are saprophytes.
2 By attacking the living tissues of organisms – that is, they are parasites.

Many of the saprophytes can destroy a wide range of man's products – paper, textiles, leather, even electrical insulation. Parasitic fungi cause serious diseases of plants. Dutch elm disease is caused by a fungus carried by beetles. A few fungi cause disease in animals; fish can be killed by fungi which attack their skin and fins. Many of the houseflies found dead on window sills have been killed by fungi whose spores they have breathed in. Athlete's foot is caused by a fungus parasite.

Moulds

Moulds will grow on a great variety of organic matter: some species will grow on dung, others on exposed foods.

Hyphae not made of separate cells joined end to end: instead, the cytoplasm has many nuclei at intervals.

Mucor (pin-mould) is a saprophyte which grows on cooked potatoes, damp bread, jam, etc. The mycelium consists of a mass of white branching threads, giving a loose fluffy appearance to the surface of food. The diagram shows the structure of a small portion of the fungus.

Like all saprophytes, *Mucor* feeds by external digestion; the hyphae secrete enzymes on to the food. After digestion, simpler foods are diffused into the mould and used to spread the growth of the mycelium. For example, the starch in bread is gradually changed to glucose. This dissolves in the liquids secreted by the mould and diffuses into the hyphae.

The food around the mould is slowly consumed. The mould cannot move and the mycelium must eventually die. Survival is possible only by reproduction. Moulds, like most fungi, produce vast numbers of spores, of which only a small number reach more food. The main method is *asexual reproduction*, an extremely efficient process which starts soon after the mycelium begins its growth. The mycelium becomes covered with reproductive structures, rather like pins sticking up from a pincushion. When ripe, the mucilage swells with water from the vertical hypha and bursts the capsule. Spores are dispersed

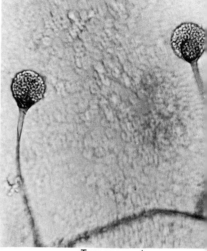
Two sporangia

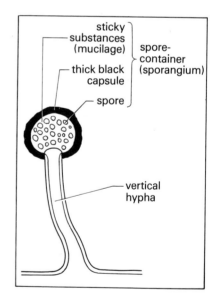

by wind or insects; they are extremely hardy. A small number may land on exposed food and germinate to give a new mycelium. *Sexual reproduction* does not occur often.

Mould fungi are important in nature because, like bacteria, they

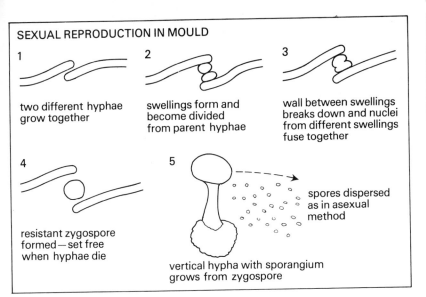
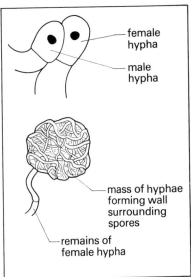

act as scavengers – moulds break down and get rid of the dead remains of other living organisms. Moulds affect man because they spoil and destroy his food. Some excrete substances which are useful medically – these substances are antibiotics. Penicillin is obtained from the mould *Penicillium*. There are many other antibiotics. When used in the correct dose in tablets, ointments, etc., they stop the growth of bacteria without any serious effect on living animal tissues. They can be used to treat a wide range of bacterial diseases in man (from boils to more serious diseases such as meningitis) and in other animals.

Mildews
Erysiphe, a parasitic fungus, causes powdery mildew, a serious disease of wheat, barley, oats and similar plants. The fungus grows its white mycelium over the surface of the leaves and stems. The host plant is seldom killed but is less able to photosynthesise and the crop is severely reduced. Suckers growing from the parasite into the epidermis cells provide anchorage and absorb food from the host's sap. Unlike *Mucor*, the

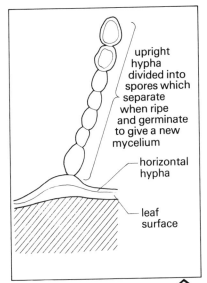

hyphae consist of single cells joined end-to-end.

The hyphae produce upright branches for asexual reproduction: these divide into spores. The spores separate when ripe, are blown away and germinate to give a new mycelium.

Sexual reproduction occurs when the ends of two hyphae grow together. The contents of the male cell (usually smaller) pass into the female and fuse together. Then the fused structure becomes surrounded by a mass of protective hyphae which grow out from the female hypha. Spores are formed inside, protected through the winter after the host has died. Next season, the covering breaks away and the spores are shot out. Each spore can develop into a new mycelium if it comes in contact with a new host.

Mushrooms and toadstools
The umbrella-shaped part above ground is called the fruiting body: it is concerned only with spore production and dispersal. The main part of the fungus lies below. In a mushroom, for example, the main mycelium is in the soil. Mushrooms are saprophytes, digesting dead remains in the soil.

Many 'toadstools' are extremely poisonous and should not be confused with mushrooms.

The reproductive structures of **bracket fungi** can be seen like shelves attached to dead trees. The mycelium is inside the tree digesting the wood. Some even attack living trees. The timber is made rotten and worthless, often before the bracket appears at the surface.

Puffballs are the reproductive structures of fungi that form their spores in a large sac. Some are edible before they become ripe. Truffles are similar – these delicacies are rooted from the soil by pigs

Fungi

Field mushrooms

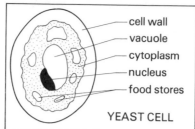

YEAST CELL

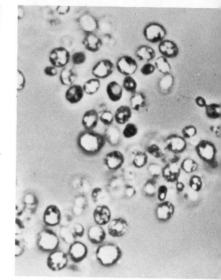

Yeast colony with budding clearly visible

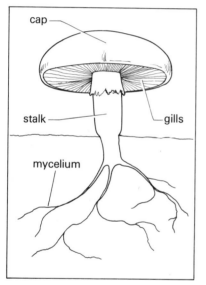

Gills are flat plates of tissue, arranged in a circle, hanging below the cap; these contain spores (as many as 10,000 million) that are flung out into the wind when ripe gills are dry. When mushroom grows rapidly upwards a ring of tissue breaks below gills. Its remains can usually be seen on stalk.

and dogs in France and Italy. Man's sense of smell is too poorly developed to detect them.

Yeast

Yeast is different from most other fungi; it is single-celled and does not form a thread-like mycelium. Each cell is roughly egg-shaped, bounded by a wall made of a special cellulose. Yeasts are found on the surface of fruits and grains, and some are present in the soil.

Yeast is able to respire without oxygen (anaerobic respiration). The cells grow very rapidly in sugary solutions. This leads to asexual reproduction by budding; one cell may produce another every two hours. The yeast cell grows a bud which becomes larger. The nucleus divides into two and one goes into the bud. Often the bud does not separate from the parent, so that when the buds have grown and reproduced, a colony of yeast cells is formed.

Sexual reproduction takes place when food is scarce. The nucleus and cytoplasm divide to give four spores. A spore may then fuse with another spore to give a new yeast cell.

USES OF YEAST

Yeast grown in sugar solutions produces alcohol and carbon dioxide as waste products. Yeast is said to *ferment* the sugar; the process is very important to man.

1 **Bread-making** Moist dough can be fermented by yeast, producing carbon dioxide. The gas trapped in the dough makes the bread 'rise' when baked.
2 **Wine-making** Grapes have yeasts on their skins. These are used to ferment the sugary juices that come out of the crushed grapes. The mixture of grape juice and yeast is called 'must': this is fermented at a moderate temperature. The alcohol produced eventually kills the yeast and the dead cells gradually settle out as 'lees'. As fermentation ceases the liquid slowly clears, is cleared of sediment, bottled and corked.
3 **Beer-making** Barley grains are used instead of grapes. The grains are first malted, that is, allowed to germinate so that some of their stored starch is broken down to sugar. The grains are then dried and ground. The malt is next mashed with warm water and boiled with hops to give the characteristic flavour and to help clear the liquid. After cooling, yeast is added. This 'wort' is then fermented in large tanks (vats).

Yeast is also used to make vinegar. Dried yeast is a valuable tonic food, as it is rich in protein and vitamin B.

Lichens

A lichen is a close alliance between a fungus and an alga. It is probably an example of symbiosis. The fungal partner obtains food from the alga (made by photosynthesis); it is not certain what, if anything, is gained by the alga.

SYMBIOSIS is a relationship between two organisms living closely together, each benefiting the other.

Most common lichens have a thin flattened structure (thallus), which is white, black or red, orange, yellow or green. Before the manufacture of artificial colours, lichens provided important dyes. Litmus, for testing acidity in chemistry, is still obtained from lichens. Many lichens are leaf-like, but others have a more brittle, crusty texture.

Lichens are extremely important in nature. They may be the first plants to grow on bare soil when all other plants have been destroyed by fire. Some grow on bare rock; an example is the orange lichen found above high water on a rocky sea-shore. Lichens can dissolve the rock to obtain anchorage: this helps the process of weathering, forming soil from rock.

Lichens are found throughout the world, even in dry deserts and cold Arctic wastes. The 'mosses' on which reindeer and caribou feed during the winter are really large patches of lichen. Occasionally, men have kept themselves alive by eating lichens. Lichens are to be found growing in most exposed situations – tree trunks, fences, etc. Lichens growing on the branches of trees (as on Dartmoor) give a good indication of unpolluted air.

Lichen growing on heather

Mosses and liverworts

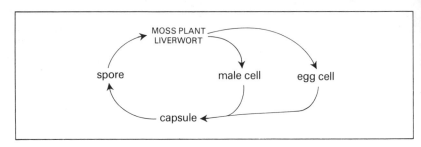

Mosses and liverworts belong to a group of green flowerless plants called the bryophytes. They are small plants which have no woody tissues. Their stems and leaves have a much simpler structure than any of the more familiar flowering plants. Their life histories are peculiar in that a fertilized egg cell does not grow into a plant like the parent. Instead, it remains attached to the parent plant and grows into a stalked capsule.

The capsule is green at first and can carry out photosynthesis, like a a normal green leaf, but it depends upon the parent plant for water and mineral salts. The capsule produces spores. From these, under suitable conditions, a new plant like the the original plant will develop. Bryophytes are only partly adapted for life on dry land. No sexual reproduction can occur unless there is external water available for the male sex cells to swim to the egg cells.

Liverworts
Liverworts usually have horizontal (prostrate) stems. The stem may be thin with many close-set leaves, also horizontal. Other kinds have a flattened branched and ribbon-like stem which also acts as a leaf. These, which are called thalloid liverworts, are a bright shiny green colour which looks almost transparent.

Female sex organs grow inside little pockets on the upper surface. Each is flask-shaped and contains a single egg cell. The male cells are contained inside little round

cavities which open directly on to the upper surface. A single capsule grows from each pocket, since only one egg cell will develop. The capsule is round. Inside it, the spores develop. The capsule also contains rod-shaped structures called elaters. These are unevenly thickened and

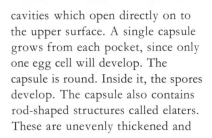

Moss growing on stone (A), Liverwort with spore capsules (B), Thalloid liverwort with vegetative propagation organs (C)

Principles of Biology

change shape as their moisture content alters in response to the changing humidity of the air. In changing shape, they flick the ripe spores from the capsule.

Mosses

A typical moss plant consists of a simple vertical green stem bearing closely-set green leaves. Each leaf, except for its slightly thickened midrib, is only one cell thick. The stem extends a short distance below the ground surface. From the underground part, the branching root-like structures (rhizoids) arise. Each rhizoid is formed from a single column of cells.

Sex organs are enclosed by the leaves at the stem tip, male and female organs being on different stems. The egg cells are contained separately inside flask-shaped organs. The male cells are developed inside cigar-shaped cases. Eggs are fertilized internally by the male cells which are able to swim by means of their cilia in the surface moisture film. Only one egg cell develops at each stem tip. It remains attached to the parent plant and grows into a spore case (capsule) carried high above the parent plant on a long stalk (seta). The capsule is often covered by the brightly coloured remains (calyptra) of the female sex organ.

By cell division many spores form inside the capsule. These are dispersed in dry weather. In moist conditions they germinate to form branching threads which turn green if exposed to light. The familiar moss plants arise as buds from these threads (protonema).

Mosses like damp shady places; some actually grow in water. A surprising number, however, survive in most unlikely places such as on bare rock, stone walls or roofs. Fertilization can take place only in wet conditions. Over 600 species of mosses are known to grow in the British Isles.

A

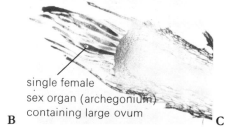

single female sex organ (archegonium) containing large ovum

B

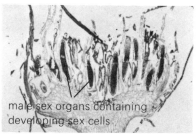

male sex organs containing developing sex cells

C

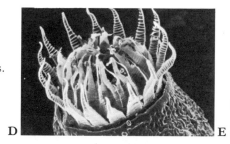

D

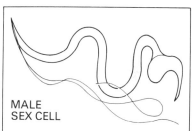

MALE SEX CELL

E

F

Side view of liverwort showing male reproductive capsules (A), Photomicrographs of female (B) and male (C) organs of mosses, Close-up of lid of moss capsule (D) – the valves open as they lose water (in dry air) and close when the air is damp, Male sex cell of moss (E), The spore cases of *Polytrichum*, a common moss with brightly coloured calyptras (F)

Mosses and liverworts

Ferns

The ferns and their allies are non-flowering plants, like the mosses and liverworts. They were the dominant land plants at the time when the coal measures were being formed about 350 million years ago. They bear green leaves some or all of which produce spores. In ferns, all the leaves produce spores. In other members of the group, only the closely-clustered leaves at the end of the stem produce the spores. Each spore gives rise to a tiny inconspicuous little plant called a prothallus. It has male and female sex organs. External water must be present

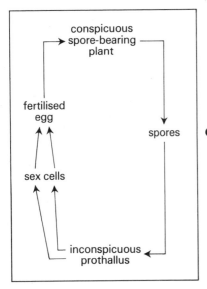

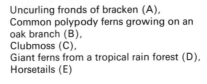

Uncurling fronds of bracken (A),
Common polypody ferns growing on an oak branch (B),
Clubmoss (C),
Giant ferns from a tropical rain forest (D),
Horsetails (E)

for sexual reproduction to take place. From a fertilized egg cell, a new spore-bearing plant will grow.

There are three main groups among the fern-like plants. The ferns usually have underground stems and large, much-divided leaves. The clubmosses and horsetails have small leaves and aerial stems which may be vertical or horizontal. Ferns are found all over Britain, the best-known species being bracken. Horsetails are widespread, being particularly common on railway embankments. Clubmosses, more rarely found, are restricted mainly to the wet hillsides in the north and west of Britain.

Principles of Biology

The male fern

The male fern is a familiar sight in gardens, woodlands, and shady hedgerows. Why it should be known as the *male* fern no one seems to know. The leaves are large (up to one metre in height). They are much divided and grow from an underground stem (rhizome). At first, the young leaf is rolled up and covered with delicate brown scales. The inner surface grows more rapidly than the outer surface and so, as it gets bigger, the leaf gradually unrolls.

The green leaves carry out photosynthesis. They also produce spores. Under each of the smallest subdivisions of the leaf (pinnules) tiny heart-shaped scales may be seen. In early summer they are green but as summer advances they turn brown. Each scale is supported on a short stalk so that it looks rather like a small umbrella. From the base of the stalk the spore cases (sporangia) grow. The spore cases open in midsummer and catapult the spores away from the parent plant.

If conditions are suitably moist, a tiny green plant called the prothallus grows from each spore. This develops male and female sex organs on its under-surface. The mobile male cells swim in the surface moisture film and fertilize the egg cells, each of which is enclosed in a flask-shaped female sex organ. Only one of the fertilized eggs will grow on a single prothallus, and it gives rise to a new fern plant.

The fern shows a life history known as 'alternation of generations'. It does not produce a plant like itself. The dominant and familiar plant is the spore-producing plant, which starts life by being dependent upon the sexually reproducing prothallus. It eventually becomes independent in the fern, though not in the moss.

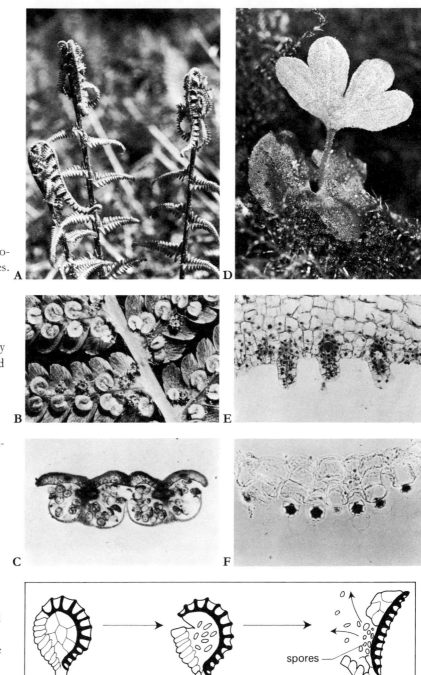

Young fronds of a male fern uncurling (A),
Underside of a male fern showing scales which cover spore cases (B),
Photomicrograph of leaf in section showing spore cases and covering scale (C),
A fern prothallus with young fern plant growing from it (D),
Section of prothallus showing female sex organs (E), and male sex organs (F)

Ferns

Conifers

Conifers are the modern descendants of a much larger group of plants which flourished in prehistoric times. Conifers dominated the landscape more than 200 million years ago. There is a similarity between conifers and reptiles – the ancestors of both groups were only partly adapted to life on land. Amphibians, the ancestors of reptiles, and ferns, the ancestors of conifers, both depended on water for reproduction. Ferns produce spores which need water for germination, and fertilization does not take place unless the male gametes can swim in water.

Conifers are *seed*-bearing plants but, unlike flowering plants (which developed millions of years later), they hold their seeds in cones instead of flowers. The cone gives some protection to the seed. Seeds, unlike spores, contain an already partly-developed young plant together with a food supply provided by the parent. They are therefore much more likely to survive and produce new plants.

Each cone has closely-packed scales spiralling around a central axis and bearing male or female spores. Male cones are usually much smaller than female cones and last for a very short time. When the cone is ripe the scales open. The male spores are pollen grains, instead of the swimming sperm cells of mosses and ferns. The bright yellow pollen is scattered by wind; it does not depend on water. Each grain has two wings, for buoyancy. Only a small portion of grains reach a female cone; vast numbers are produced and wasted. The female spores are attached to the upper surfaces of the scales in the female cone. Any pollen grains that land near these spores may grow pollen-tubes to fertilize the egg cells inside.

After pollination, the scales of the female cone close again to protect the seeds as they develop inside fertilized female spores. When the

Mature male pine cones in May (A), Female pine cone (three years old) (B), Yew twigs – the seeds are surrounded by bright red flesh (C)

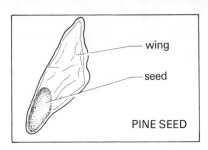

PINE SEED

scales later re-open, the seeds can be shaken out. Reproduction in some conifers, such as pine, takes several years. Many conifer seeds have wings to assist dispersal.

In yew (where the female cone produces one seed only, at its tip), the seed develops a bright red fleshy cover to attract birds, which disperse the seeds.

The seeds of all conifers lie exposed on the surface of the cone scales: this contrasts with flowering plants where they are enclosed and protected by an ovary wall.

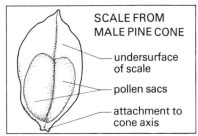

SCALE FROM MALE PINE CONE
- undersurface of scale
- pollen sacs
- attachment to cone axis

Ripe pollen grain of Scot's pine (A), Giant conifers (B)

Conifers are an extremely successful group of plants. In Britain they have been extensively planted by the Forestry Commission, notably on mountain slopes where the climate in exposed situations is severe and where the roots help to prevent soil erosion by wind and rain.

Conifers grow much faster than trees such as oak and ash (which have flowers, not cones). But the xylem tubes of conifers have walls that are much less heavily thickened – conifer wood is *softwood*, used for making newspaper pulp and for the interiors of buildings. Whitewood furniture is made from Norway spruce and young trees are greatly in demand throughout the world at Christmas. Deal comes mostly from pine.

Conifers constitute nearly half the world's forests. Typically, they grow where life is too difficult for other plants, especially where there is water shortage – in the Arctic and at high altitudes where the soil-water is frequently frozen and therefore unavailable, and in sandy deserts. The shapes of their leaves, generally long and like needles, or scaly, gives them a relatively small surface area. This restricts water loss by transpiration. Most conifers are evergreen and keep their leaves at times when water shortage is most acute. The European larch is an exception.

Most of the world's largest trees are conifers. The largest conifers are the Californian redwoods, some well over 100 metres high. Gigantic conifers are found elsewhere – in South America and New Zealand – with diameters of ten metres or more.

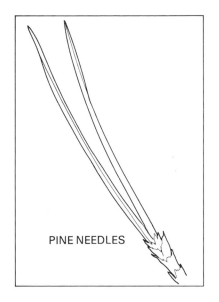

PINE NEEDLES

Flowering plants

In many ways, flowering plants are to the plant kingdom what mammals are to the animal kingdom. They are the dominant land-living organisms of the present day. They show important advances on conifers. Instead of having exposed seeds on cones, flowering plants enclose their seeds in a protective ovary. Cones are replaced by special reproductive organs, called flowers. In flowering plants, the pollen grains are not necessarily carried to the female parts by wind; they may be carried by insects instead. This is a more reliable method if the flower has developed structures attractive to insects.

Flowering plants have three basic types of organ – root, stem, leaf – each with its own functions. At certain times, flowers are developed for sexual reproduction.

Healthy roots grow continually from the division of cells just behind the root tip. As the tip is pushed forward it is protected by a root cap of tough cells, constantly rubbed away by friction with the soil. Root-hairs are found always a short distance behind the tip; they live for a few days only and are rubbed off. They are continuously replaced by new hairs growing immediately behind the advancing tip.

The strongest part of a root is the xylem at the centre. This arrangement probably gives the greatest flexibility (to grow around large soil particles and stones) combined with greatest resistance to forces that could pull the root out of the ground.

Some plants (grasses, for example) develop a fibrous root system – there are many roots, all about equal size.

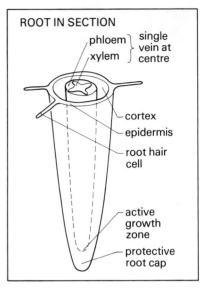

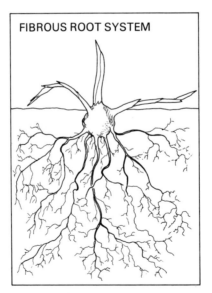
FIBROUS ROOT SYSTEM

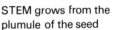

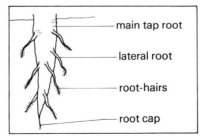

Xylem has tubes with thick walls – for support, and transport of water and salts up to the stem.
Phloem has tubes with thin walls – carry dilute solution of foods manufactured in leaves.
Epidermis is a single layer of protective cells.
Root-hairs absorb water and salts from soil.

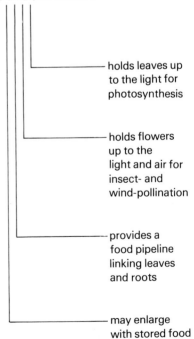

ROOT grows from the radicle of the seed
— anchors plant in ground
— absorbs water and salt from soil
— may enlarge with stored food during resting season

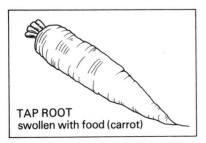
TAP ROOT swollen with food (carrot)

66 **Principles of Biology**

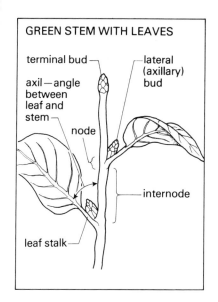

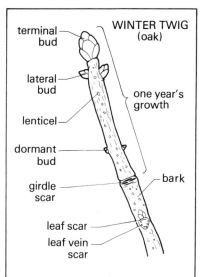

woody plants only the inner parts are green – the outer structures are hard and brown, forming protective bud scales. These scales fall away as the bud opens. Some buds grow out into branch stems with their own leaves; others produce flowers and then stop growing. A few buds are *dormant* and may never develop. They produce side branches only when the terminal bud is damaged. Gardeners make use of this, pinching out terminal buds to make bushy plants.

Stems have several veins arranged in a circle near the outside. These veins join at ground level forming the single central vein of the root. Stems have to resist bending by the wind and when animals brush past. Bending forces are stronger towards the outside; this is where stems must be strongest. The xylem in the veins forms strong cables along the outer part of the stem.

Stems have leaves and buds. Some stems are soft and green and have a brief life (usually one growing season). These plants are *herbs*. Other stems are hard and brownish, covered by a tough bark – they live for much longer and are *woody*; these are trees and shrubs. (Shrubs are generally smaller than trees; with almost equal branches, instead of one large trunk with branches.) Woody plants have a resting, dormant period – in Britain, this is during the cold winter weather. Winter twigs show many interesting markings.

Dormant bud does not develop unless other buds are damaged. Lenticels are microscopic craters allowing entry and exit of gases. Bark (tough woody covering) prevents water loss and entry of disease; protects from cold. Leaf scar shows where leaf was attached and position of veins.
Girdle scar marks position of scales of previous year's terminal bud.

Buds are very young compact shoots; the stem is extremely short and bears undeveloped leaf-like structures wrapped tightly around it. In buds like the Brussels sprout these are green, but in the buds of

Epidermis – single layer of protective cells.
Cortex – packing and food-storing cells.
Phloem has tubes with thin walls – carry manufactured food away from leaves.
Xylem has tubes with thick walls – for support and to transport water and salts up from roots.

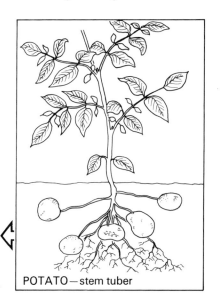

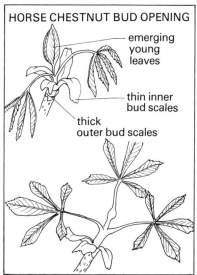

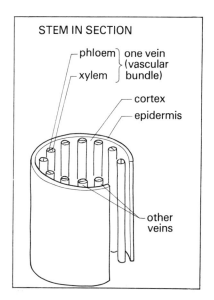

Flowering plants 67

LEAF
leaves, like flowers, develop from buds on stems

- provides a large flat surface to trap light for food manufacture (photosynthesis)
- has pores and passageways for movement of gases
- may enlarge with stored food during resting season

Leaves can be *deciduous* – shed regularly once a year at the start of the resting season (in Britain, in the autumn) or *evergreen* – remaining on the plant during the resting season and shed at no definite time.

The sizes and shapes of leaves are different in each species of plant. Leaves are *simple* if there is only a single leaf blade, and *compound* if the leaf blade is divided into leaflets all showing one attachment to the stem. The edges (margins) also vary – some are smooth (apple), others wavy (oak) and others deeply toothed

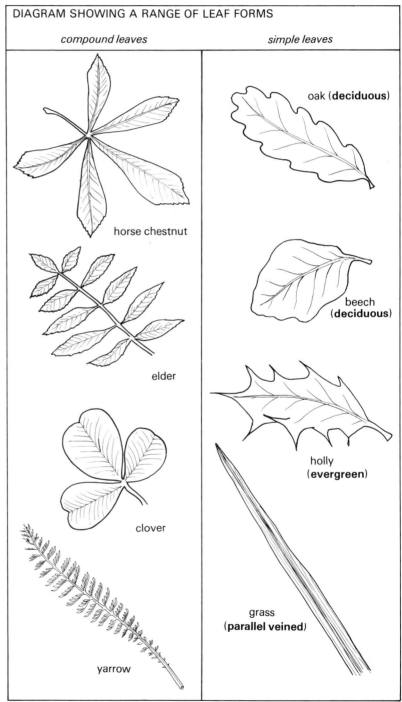

DIAGRAM SHOWING A RANGE OF LEAF FORMS

compound leaves: horse chestnut, elder, clover, yarrow

simple leaves: oak (**deciduous**), beech (**deciduous**), holly (**evergreen**), grass (**parallel veined**)

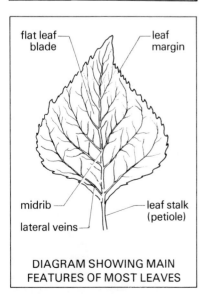

BULB IN SECTION
- fleshy leaves swollen with stored food
- outer covering of protective papery leaves

DIAGRAM SHOWING MAIN FEATURES OF MOST LEAVES
- flat leaf blade
- leaf margin
- midrib
- leaf stalk (petiole)
- lateral veins

Principles of Biology

(nettle). The leaves of grasses and of plants such as the lily family are long and strap-shaped with the veins *parallel*; otherwise leaves are usually shorter and broader with the veins forming a network (*net-veined*).

main vein, bringing water and salts from root to leaf for food manufacture, and carrying manufactured food away.
Upper and lower epidermis – single layer of protective cells covered on outside with waxy cuticle to reduce water loss. Cuticle very thick in evergreen leaves, giving glossy appearance.
Stoma – pore in epidermis leading into system of air spaces. Allows entry and exit of gases for respiration and photosynthesis.
Spongy layer – rounded cells with relatively few chloroplasts.
Palisade layer – columns of cells containing many chloroplasts, for photosynthesis.

Life cycles of flowering plants in Britain

Annuals live for one season only. They flower and die away, leaving only the seeds to survive the winter and germinate the following spring. The life cycle of an annual is shown in the diagram below.

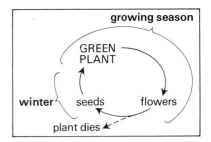

Biennials live for two seasons only. In first season, produce green parts but no flowers. Then die away, leaving food storage organ only, protected below ground during winter. In second season, stored food used to grow new green plant which flowers. Whole plant then dies leaving only seeds to survive winter and germinate next spring.

Perennials live indefinitely. Flowers produced and seeds dispersed each season, except perhaps the first season. Main plant survives each winter *either* below ground in the form of a food storage organ *or* above ground protected by bark.

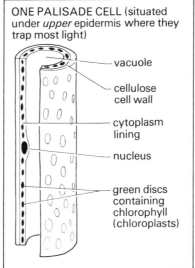

Epidermis is a tightly-fitting jigsaw of cells forming a complete protective covering.

Guard cells can swell with water or collapse to open or close the stoma.

Midrib supports leaf blade and holds it flat to the light. Midrib contains

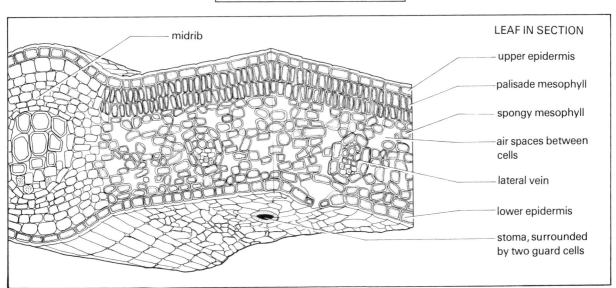

Flowering plants

Part 4
The organism: in action

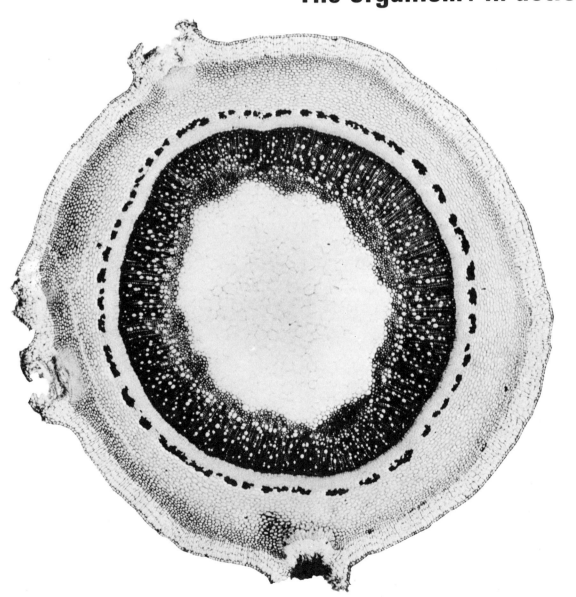

Nutrition

Nutrition is concerned with feeding: it occurs in all living organisms. They
either take in simple materials and manufacture these into food
or take in ready-made foods and break these down into a form that the organism can use.

There are several variations within these two types.

Food is really stored energy from the sun. Living organisms release this energy for their own purposes. Energy is needed to produce essential substances and also when muscles contract for movement and when messages are sent along nerves. Some energy is released as heat – this is important for keeping a constant body temperature in birds and mammals. Growth takes place by making living cells larger and by making more of them. In order to do this, food has to be altered in chemical processes and at the same time other food has to provide the energy for these processes.

Photosynthesis – the nutrition of green plants

Photosynthesis is the main method of food manufacture in the living world. As in all manufacturing processes, a supply of energy is required. In photosynthesis, this comes from the sun.

Photosynthesis is important for two reasons. It produces almost all the world's food, for both plants and animals. Every food eaten by an animal can eventually be traced back to photosynthesis.

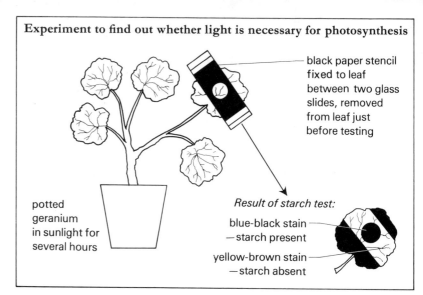

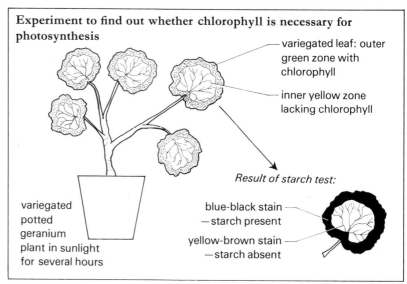

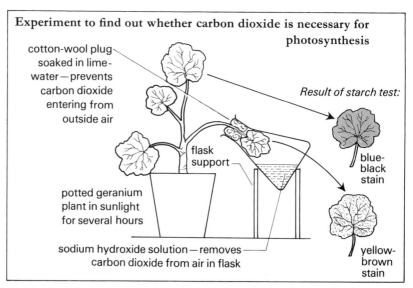

Also, the balance between oxygen and carbon dioxide in the atmosphere is maintained by photosynthesis. All living organisms breathe out a suffocating gas, carbon dioxide.

In sunlight, green plants use this gas for photosynthesis. They exchange it for equal amounts of oxygen (as shown in the equation), which all but a few organisms need for respiration.

Photosynthesis takes place only in living cells in light. Two simple substances, carbon dioxide and water, are built up into simple sugars with oxygen given off as a waste product. Green plants are able to feed by photosynthesis because they contain the green pigment chlorophyll.

$6CO_2$ **carbon dioxide**

+ **is combined with**

$6H_2O$ **water**

in the presence of sunlight and chlorophyll — **forming**

$C_6H_{12}O_6$ **glucose**

+ **and**

$6O_2$ **oxygen gas**

Every green part of a plant photosynthesises, but leaves are the main organs for photosynthesis as they generally contain most chlorophyll. In some simple plants (some seaweeds), brown and red pigments are present which absorb light for photosynthesis: this is to make better use of the light, mainly blue, available in deeper waters. But most plants need four essentials for photosynthesis – sunlight, chlorophyll, carbon dioxide, water.

Usually, as the amount of glucose builds up in the plant during the day, it is converted to starch for storage in the cells. At night, when photosynthesis does not take place, starch is changed back to sugars and distributed around the plant. Starch formation is used in experiments to test the conditions for photosynthesis. If starch is formed, then photosynthesis has taken place. In all these experiments, though, it is necessary to remove the starch already present in the leaves before the experiment begins – this is done by placing the plants in the dark for forty-eight hours. At the end of the experiments, leaves are removed from the plants and iodine-tested for starch using the following method:
1 Dip leaf in boiling water to kill.
2 Place leaf in methylated spirits (warmed indirectly over a water bath) to remove chlorophyll.
3 Wash leaf and soak in iodine solution.

The production of oxygen
Green plants give off oxygen gas as a waste product from photosynthesis. It diffuses out through the pores (stomata) of land plants. Bubbles collect on the leaves of water plants. Some of the oxygen rises to the surface and some dissolves in the water where it is used by organisms such as fish. For this reason, submerged weeds are an essential feature of an aquarium. In very bright light, so many oxygen bubbles appear that they can be tested in the following experiment:

Experiment to find out what gas is given off during photosynthesis
A control experiment is carried out at the same time – exactly similar apparatus as shown in the diagram but kept in darkness.

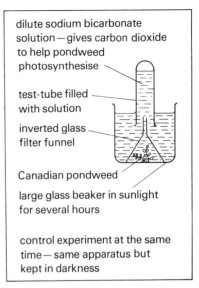

Bubbles of colourless gas rise in the tube and displace the solution. The gas is tested either with a glowing splint (which ignites) or by adding alkaline pyrogallol (which absorbs the gas so that liquid returns to the test-tube). Both tests show that the gas is oxygen.

After photosynthesis
The sugar made by photosynthesis remains in the leaf or is carried by the phloem tubes in the veins to other parts of the plant. Some is used to make cellulose, a substance rather like starch, needed for building new cell walls. But green plants do not feed themselves by photosynthesis only. Sugars and starches are mainly energy-givers, but not growth substances. For growth, plants absorb mineral salts – from the soil, for example. Foods made by photosynthesis then react with these salts, producing proteins (for growth) and other essential substances. Chemical reactions of this sort need energy, which comes from the breakdown of some of the sugars made by photosynthesis. In a leaf, then, some of the sugars are being broken down while at the same time others are being built up.

Nutrition 73

Results of water culture experiment

The importance of salts for healthy plant growth can be shown by using different *water cultures* – solutions of different salts in water – to find out how growth is affected when certain minerals are absent.

Water cultures
Glass jars filled with different solutions and covered with dark paper to prevent growth of algae. Each jar fitted with split cork so that similar healthy seedlings can be placed in jars with roots in solutions. Each cork has two tubes which pass down into the solution so that air for root respiration can be blown through regularly. Experiment left in light, cool place and examined over several weeks.

The photograph shows that only seedlings grown in balanced salt solutions thrive. Seedlings cultured in distilled water soon cease growing. So do seedlings in cultures lacking nitrates – plants take in nitrates and use the nitrogen to make growth substances (all proteins contain nitrogen). Magnesium and iron salts are needed for plants to make chlorophyll. Leaves turn yellow when plants are grown in solutions lacking these elements.

The nutrition of mammals
Mammals, unlike green plants, cannot manufacture their own food. Like most other animals, they take in ready-made food (*ingestion*) and break the food into smaller units (*digestion*) which can be taken up from the digestive system (*absorption*). These small units are then built up into the animal's own tissues (*assimilation*) and any undigested food is passed out of the body (*egestion*). The food of animals ultimately comes from the food manufactured by plants.

The food ingested by mammals contains large molecules. These molecules are too complex to pass through the intestine wall into the bloodstream. Digestion is necessary to make the food molecules simpler and some have to be made capable of dissolving in the blood plasma.

Food is processed from the time it enters the mouth. Some digestion is mechanical – food is sliced, torn and ground by the teeth and squeezed by muscles in the wall of the digestive system. Chemicals, called enzymes, secreted by digestive glands, are also needed.

AN ENZYME –
  speeds the process of digestion
  is effective at normal body temperature
  acts only in conditions of a certain acidity or alkalinity
  digests one type of food only

Food substances

Foods are required in bulk by living organisms for two main purposes – to provide building blocks for growth and as energy sources. Energy foods are called calorific because the energy they give is measured in calories or joules. Sometimes energy foods are used for growth, and growth foods are broken down to release energy. Besides carbohydrates, fats and proteins in bulk, organisms need small quantities of accessory food substances – vitamins and mineral salts.

Vitamins – essential in small quantity for the normal functioning of the body. Plants and animals can make vitamins, but animals usually must take in some vitamins with their bulk foods. Absence of vitamins causes serious deficiency diseases. Some vitamins dissolve in water; others dissolve only in fats – these are obtained by eating fatty foods such as butter and margarine.

FAT-SOLUBLE VITAMINS

Vitamin A
Lack causes night-blindness (poor vision in dim light) and damage to the living membranes of eyes, nose and throat.
From fatty foods (egg yolk, margarine, fish liver oils, etc.), dairy produce and also from green vegetables.

Vitamin D
Lack causes rickets: bones do not fully harden and are easily bent or broken, leading to deformities (particularly in the young, where growth is rapid).
From fatty foods and dairy produce.
Also made naturally in the body by the action of sunlight on skin.

Vitamin E
Lack may affect reproduction: sperm may be infertile and embryos may fail to develop.
From fresh green vegetables, whole cereal grains, eggs and dairy produce.

WATER-SOLUBLE VITAMINS

Vitamin B
(a group of vitamin substances)
Lack causes eventually fatal diseases of nervous and digestive systems: beri-beri (muscular weakness and paralysis) and pellagra (cracking and flaking skin).
From whole cereal grains, liver, meat, eggs, green vegetables, peas, beans, yeast.

Vitamin C
Lack causes scurvy, eventually fatal: internal bleeding and bruising and also bleeding from nose and gums – wounds heal more slowly.
From many undamaged fruits (oranges, lemons, grapefruits, blackcurrants, tomatoes, rosehips) and from fresh green vegetables and potatoes.
Unstable – soon destroyed by exposure to sunlight and air.
Scurvy was a common disease during long sea voyages when fresh fruit and vegetables were not available.

Mineral salts – these are essential for both plants and animals. They are taken in in relatively small amounts. 'Trace elements' (such as copper and zinc) required by many plants must be absorbed in extremely small quantities – larger amounts would be poisonous.
Some of the most important elements are:

NITROGEN and SULPHUR
For the formation of growth substances such as proteins.
Green plants absorb these elements as nitrates and sulphates: animals take in and digest proteins.

PHOSPHORUS
For the formation of growth substances and for the hardening of bones and teeth.
Green plants absorb phosphates: animals take in and digest protein foods.

IRON
For the manufacture of chlorophyll and formation of blood oxygen-carrier haemoglobin.
Green plants absorb iron salts. Animals take in iron compounds in foods (e.g. liver, watercress).

MAGNESIUM
For the manufacture of chlorophyll and hardening of bones and teeth.
Plants absorb magnesium salts: animals take in cereals and vegetables.

IODINE
For the secretion of thyroid hormone.
Obtained from sea foods.

FLUORINE
For the hardening of bone and tooth enamel.
Obtained from milk and drinking water.

Nutrition

Many enzymes can be used to investigate enzyme action, but the most convenient is the amylase ptyalin in human saliva. Amylase enzymes digest starch, and at the start of the experiment some starch solution is tested with iodine solution. The remainder of the starch solution is mixed with saliva from the mouth and treated as shown in the diagram.

Experiment to find out whether saliva contains an enzyme

Boiling destroys enzymes and provides a control. Further test tubes can be used, some with

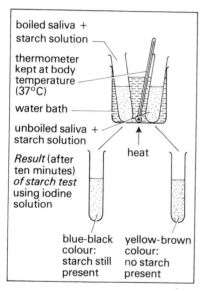

acid or alkali added to the starch-saliva mixture; these should inactivate the enzyme so that the starch is not digested.

Balanced diets
Mammals need a regular intake of carbohydrates, fats and proteins together with small quantities of vitamins and minerals. Larger mammals need a larger food intake. Nevertheless, very small mammals lose a great deal of heat energy over their relatively large surfaces and so their food requirements are disproportionately large.

Water has to be taken in also.

Much of the body is fluid and many living processes take place only in solution. Water is continually being lost so that the amount of water required varies with circumstances – very active mammals, and those living in hot dry regions, will require greater amounts of water.

Mammals also require roughage. This is food which remains largely undigested and provides bulk for the intestine muscles to squeeze on – this helps to push the food onwards. In man, cellulose cannot be digested; vegetable material, with its high proportion of plant cell walls, constitutes roughage.

Carbohydrates, fats and proteins must be ingested in the correct ratio to give a balance between calorific and other foods. A man employed in heavy manual work will require more food each day than a bank clerk, and he will need relatively more calorific food. If an inactive man eats too much, and, in particular, too many high-calorific foods, the excess will turn to fat and he will become overweight.

Energy output per minute of average-sized man at different activities
Sitting reading 1½ kcal
Walking at 2 m.p.h. 2 kcal
Papering a wall 3 kcal
Repairing a car 4½ kcal
Walking at 4 m.p.h. 6 kcal
Playing football 7½ kcal
Walking upstairs 10 kcal

Balanced meals
When young and actively-growing, mammals need relatively larger amounts of food, but again, the main classes of food substances must be present in suitable proportions. The lunch shown in the photograph is suitable for a schoolboy or girl. The particular nutritional value of the individual items in each course is given.

Dealing with food

Food is processed as it passes along a tube called the *alimentary canal*. Each region of the tube has its own appearance and special functions. There are also glands – organs which secrete digestive fluids and pass them along small tubes (ducts) into the alimentary canal at various points.

Teeth
Teeth are essential equipment for dealing with food at the start of digestion. In flesh-eaters

A balanced meal

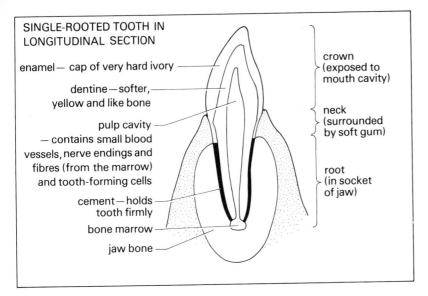

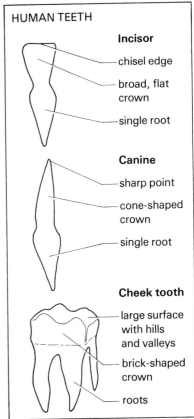

(carnivores) they can hold and kill prey. Teeth near the front of the jaws may be used to crop grass (in cattle and sheep) or for shredding hard plant materials (in rabbits), making food small enough to enter the mouth. Teeth are also important for *mastication* – breaking food into pieces small enough to swallow. Chewing a large piece of food produces many smaller pieces with a much larger surface area. Digestive juices can then attack the food more effectively.

Teeth are fixed by their roots into sockets along the jaw bones. They move slightly in their sockets during use. The largest teeth have more than one root. Each of the three main types of tooth – incisors, canines and cheek teeth (premolars and molars) – has its own special function, depending on the animal's diet. Many mammals have a milk set followed by a permanent set. Human babies cut their first set of twenty teeth (molars are absent) several months after birth. These are replaced by a permanent set from about five years old. This set has molars, but the back molars ('wisdom teeth') may not be cut until the jaws have grown longer in the late teens, or even later. Other mammals, e.g. rabbits, have their milk teeth at birth.

In man, the teeth can deal with most types of food. The permanent set has eight incisors, four canines, eight premolars and twelve molars – sixteen teeth in the upper and sixteen teeth in the lower jaw. The upper and lower incisors slide past each other, slicing the food so that it can enter the mouth cavity. The canines assist in this. The cheek teeth finally grind the food.

Rabbits have no canines; instead, the cheeks grow inwards between incisors and premolars to form a pad which helps to pass food backwards. The teeth have open roots – the opening into the pulp cavity remains wide throughout

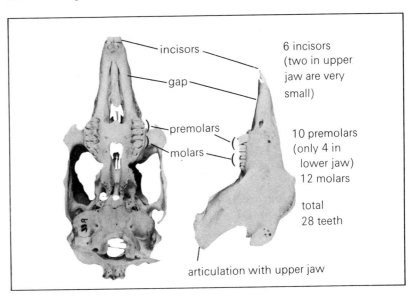

Rabbit's jaws – upper jaw (from below) and lower jaw (side view)

Nutrition 77

life, so that the teeth receive a good food supply and continue growing. The large incisors have thicker enamel on the front surface: the back and sides wear away quickly leaving a hard ridge in front. These sharp edges on the upper and lower incisors meet like chisels for gnawing. The continual growth of teeth balances the amount of wear. Wear also produces a pattern of enamel ridges on the cheek teeth. The attachment of upper to lower jaw is loose, allowing the lower jaw to rotate and move up and down. The ridges make a 'grinding mill' for vegetation as the jaw rotates – wear is balanced by new growth.

The teeth of grazing animals also have open roots and grow continually. In a sheep, the incisors and canines look alike. They bite against a horny pad at the front of the upper jaw. The grinding surfaces of the cheek teeth have raised enamel ridges. The very loose articulation of the lower jaw allows the teeth to rotate over the harsh vegetation. This millstone action is extremely effective for 'chewing the cud'.

In carnivores, the roots of the teeth are closed and the teeth stop growing when fully cut. A dog uses its incisors to hold prey and strip small pieces of meat, especially meat close to the bone; they appear as a threatening snarl when the lips are bared. They are more pointed than man's but can easily be distinguished from the long stabbing canines, which hold and kill prey and rip flesh. Each canine has a relatively enormous root extending far into the jaw. Most of the cheek teeth have rather small crowns. They form hillocks, effective for crushing. More difficult parts of a carcase are dealt with by a pair of cheek teeth called *carnassials*, more than halfway along each jaw. These are the largest cheek teeth, with their

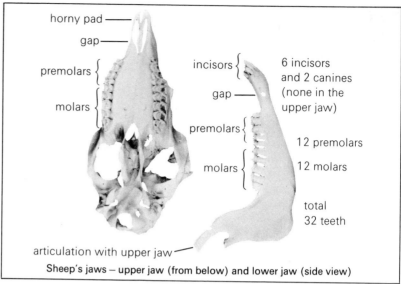
Sheep's jaws – upper jaw (from below) and lower jaw (side view)

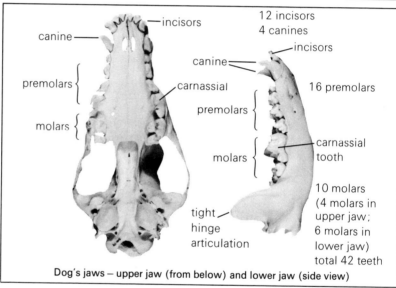
Dog's jaws – upper jaw (from below) and lower jaw (side view)

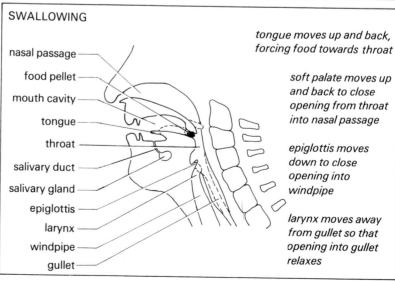

SWALLOWING

tongue moves up and back, forcing food towards throat

soft palate moves up and back to close opening from throat into nasal passage

epiglottis moves down to close opening into windpipe

larynx moves away from gullet so that opening into gullet relaxes

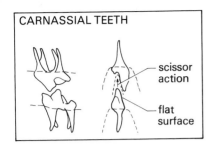

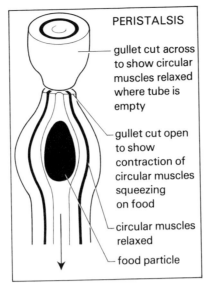

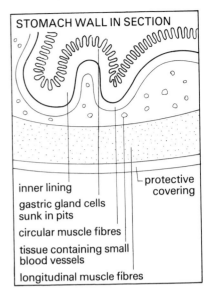

hillocks much flatter on one surface. When the jaws close, the flat blade of the upper carnassial slides past outside the flat blade of the lower carnassial. This scissor action is suited to slicing bone, tendon, etc. The articulation between the jaws forms a very tight hinge joint. This, with the powerful jaw muscles, gives the characteristic up-and-down movement, the 'snap' of the jaws.

Digestion
Food being chewed is moistened by saliva. This continually flows down small ducts into the mouth cavity from salivary glands in the cheeks and under the tongue. Sight, smell and taste of palatable food stimulates increased flow of saliva. A man may secrete more than one litre a day. Saliva is mainly water with a slimy substance (mucus) for lubrication and to hold the food together when swallowing. Dissolved salts in saliva provide the correct (nearly neutral) medium for salivary amylase (ptyalin): this enzyme, in the saliva of many mammals, digests starch to soluble malt sugar (maltose).

The tongue moves the food around, pressing pieces between the teeth and mixing them with saliva. Small particles, held in mucus, are collected by the tongue into a pellet (bolus), ready for swallowing. The actions which result in food passing across the throat (pharynx) into the gullet (oesophagus) are shown in the diagram.

After swallowing, food moves without conscious control, pushed along by circular muscles in the gullet wall. This is *peristalsis*. Circular muscles contract just behind the food and squeeze it along into the next part, where the muscles are relaxed. Peristalsis is rather like waves or ripples slowly moving over a pond.

The gullet is a narrow tube with its inner lining folded inwards when no food is present. It carries food down the neck, through the chest and diaphragm and into the abdomen. Liquids trickle down very quickly but solid food takes longer to reach the stomach.

The stomach is like a bag for storing food. It expands as it fills, and collapses when empty. Without it, we should have to eat more often. The stomach lies across the upper abdomen towards the left side, and has a capacity of about one litre in man. Food may remain there for several hours.

The exit from the stomach into the first part of the intestine (duodenum) is controlled by a ring of muscle, the pyloric sphincter. When tightly contracted, it keeps food in the stomach, like an elastic band around the neck of a bag. The duodenum is a quite narrow tube and the sphincter prevents congestion by allowing only small

Deep pits opening into the stomach contain gland cells which secrete gastric juice on to the food. Gastric juice contains hydrochloric acid and two enzymes. The strong acid stops the digestion of starch (by ptyalin) and kills bacteria that would decay food at the high temperature of the stomach – food is always contaminated by putrefying bacteria. The gastric enzymes act most effectively in the acid conditions. Rennin *makes milk easier to digest by changing the milk proteins into curds, a semi-solid form.* Pepsin *begins the digestion of proteins. (Rennin is particularly abundant in the stomachs of very young mammals.)*

Muscle layers contract when food is present to churn, mix and break up food. The heat of the stomach helps to melt solid fats such as butter and margarine. Substances that do not need digesting (alcohol and simple sugars such as glucose) are absorbed through the stomach walls into the blood.

quantities of the acid food mixture (called chyme) to pass through. The stomach muscles are able to squeeze food through a small opening into the duodenum when the sphincter relaxes. Contraction of the sphincter closes the opening. Food is forced into the

Nutrition

duodenum in a series of jets as the muscle contracts and relaxes.

The acid food stimulates cells in the wall of the duodenum to pour hormones into the blood. These are carried to the liver and pancreas, causing the secretion of *bile* (from the gall bladder) and *pancreatic juice*. This is a good example of chemical co-ordination in the body; it makes sure that digestive juices are poured into the duodenum only when food is ready to be digested there. It prevents wastage of important body fluids. The hormone *secretin* is produced by the duodenum.

Bile, which flows down the bile duct, is made in the liver and stored in the gall bladder. It is an amber-coloured liquid containing pigments formed from the destruction of worn-out red corpuscles – it is an excretion as well as a secretion. It contains bile salts but no enzymes. The salts act on oily substances in the duodenum, separating them into small floating droplets. This is called emulsification and makes digestion easier by increasing their surface area. The enzymes which complete digestion in the intestine work most effectively in alkaline surroundings: the bile salts react with the acid food from the stomach and provide these surroundings. (The alkaline food mixture is called chyle.)

Pancreatic juice contains enzymes and enters the duodenum at its U-shaped bend. One of its enzymes is an amylase and finishes the work of saliva in digesting starch to malt sugar. Another begins digesting emulsified fats, splitting them into fatty acids and glycerine.

Beyond the duodenum, the ileum completes the digestion of most foods. Glands in this region of the intestine secrete *intestinal juice*, containing several enzymes. All compound sugars, such as malt, cane and milk sugars, are converted to simple sugars such as glucose. Enzymes complete the digestion of proteins, begun in the stomach. All proteins become amino-acids.

Absorption

The duodenum and ileum are not simple tubes. 'Fingers' project from their inner surfaces; these *villi* greatly increase the surface in contact with digested food. Food can be absorbed at a faster rate. Each villus contains blood vessels and a lacteal.

The ileum is the longest tube in the alimentary canal – several metres long in man. Because of its length and villi, most of the digested food has been absorbed before the next part (colon) is reached. In rabbits, an important part of the vegetable diet – the *cellulose* of plant cell walls – has not yet been digested. Here, food enters the caecum before it passes into the colon. The caecum is more like a sac than a tube and is closed at one end by a narrow portion, the appendix. Food has to enter and leave at the same end. Bacteria living in the caecum secrete enzymes that digest the cellulose – the rabbit itself is unable to pro-

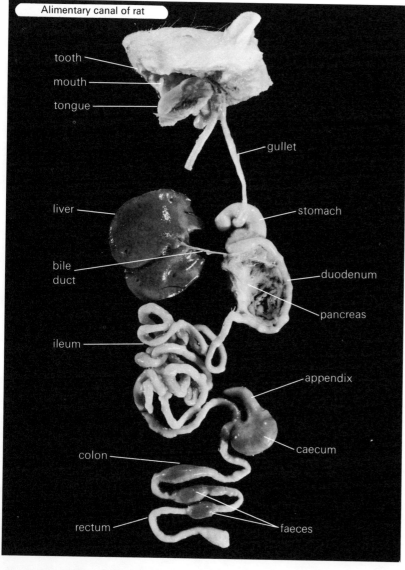

Alimentary canal of rat

duce these enzymes – forming simpler substances which the bacteria themselves use. However, the enzymes are so effective that much digested cellulose is left over for the rabbit to absorb. Rabbit and bacteria, then, are two different organisms assisting each other (*symbiosis*). The rabbit finds food for the bacteria which the bacteria then digest for the rabbit.

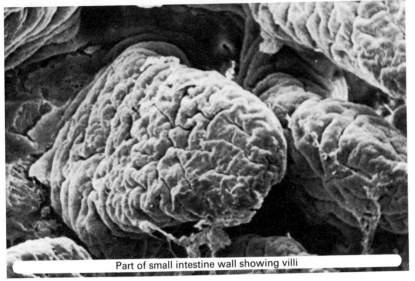
Part of small intestine wall showing villi

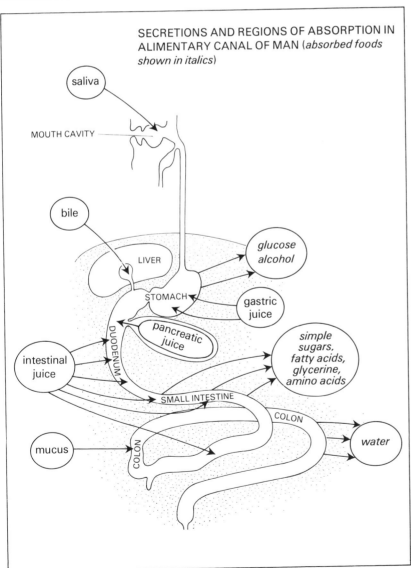
SECRETIONS AND REGIONS OF ABSORPTION IN ALIMENTARY CANAL OF MAN (*absorbed foods shown in italics*)

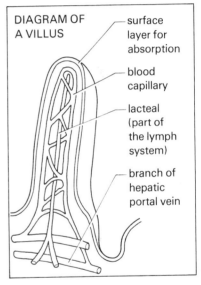
DIAGRAM OF A VILLUS

Digested foods are absorbed through the villus wall.
Simple molecules such as glucose and amino-acids pass into the blood capillaries.
Fatty substances pass into the lacteal.

Removal of waste food (*egestion*)
Liquid material enters the colon, mainly undigested food and the animal's own digestive secretions. The walls of the colon extract as much water as possible, otherwise large amounts of valuable fluid would leave the body. The food

Nutrition

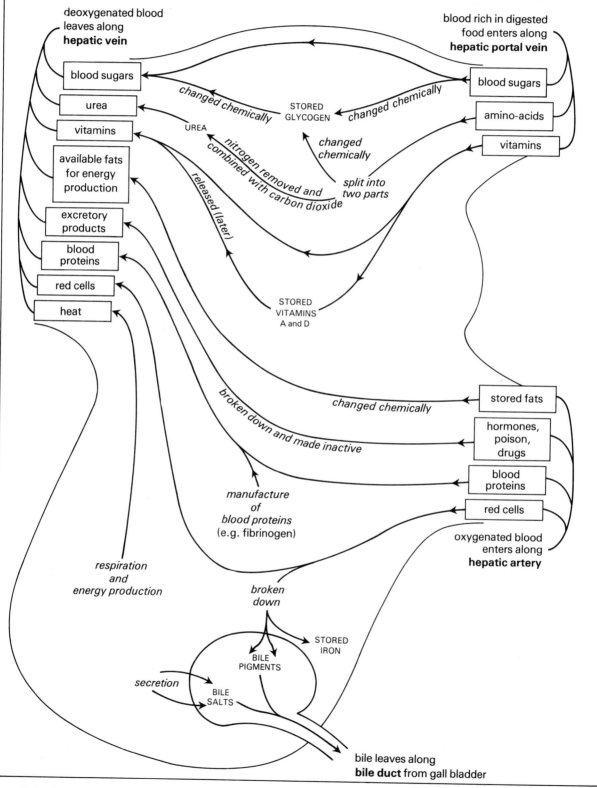

Principles of Biology

remains become semi-solid (*faeces*), coloured by the excreted bile pigments. Both the colon and the rectum (the final piece of tubing) secrete mucus for lubrication. The rectum stores the faeces until they are egested – forced out of the anus by muscle contraction.

The use of food (assimilation)
Digested foods are carried away from the intestine in the blood and lymph systems. The blood capillaries of the villi feed glucose and amino-acids into the hepatic portal vein, a direct link between alimentary canal and liver. The liver keeps the amount of blood sugar constant. All the sugar not immediately to be used by the body is removed from the blood as it flows through the liver. It is converted to a storage product, *glycogen* (sometimes called 'animal starch'). Most glycogen is stored in the liver itself but some is stored in muscle tissue. It remains there until the level of blood sugar falls, as it must do when no more sugar is being absorbed from the intestine but is still being used by the cells. Then, glycogen is changed back to sugar.

Amino-acids cannot be stored. Those needed immediately for making proteins pass through the liver. The remaining amino-acids are chemically changed by the liver – their nitrogen is taken away and the amino-acids are converted to carbohydrate or fat. The liver changes the waste nitrogen into urea, by combining it with carbon dioxide from respiration. Breakdown of unwanted amino-acids in the liver accounts for much of the urea excreted, particularly by adults.

Fatty substances collected by the lacteals of the villi bypass the liver in the lymph system. They are then emptied into a large vein at the base of the neck and distributed by the blood. Cells in certain parts of the body – under the skin, for example – can absorb fatty materials to form fat stores. Any amount can be stored, so that overeating leads to overweight.

Eventually the digested foods enter the cells to be used. Amino-acids are built up into essential cell constituents for growth. Some are used (by gland cells) for making important secretions. Fatty substances are again needed for building cell components but many are broken down as fuels.

Simple sugars such as glucose diffuse into the cells and are also respired to produce energy.

Transport

A complex organisation works efficiently only where its different parts are linked by efficient communications. This is true for business organisations, cities and countries. In Britain, distant parts are linked by lines of communication for both goods and information. Goods are carried (between London and Glasgow, say) along road, rail and air links; information, in the form of letters, newspapers, etc. also passes along these links. Urgent messages and information are carried much faster, as electrical impulses along radio and telegraph networks.

Equally essential are the lines of communication within complex living organisms such as flowering plants and mammals. Again, different regions of the body may be far apart – the brain of a giraffe is some distance from the feet, and the top of a mature tree is remote from the root ends. In a mammal, the nervous system provides a fast system of electrical communication. The blood system is used for slower transport: it carries raw materials such as oxygen and other goods (foods), but it also carries instructions (chemical messages, in the form of hormones). No cell is far from the blood.

Plants also have lines of communication, but rather simpler than those of animals.

Transport in mammals – the blood system

The blood of mammals circulates in one direction only, along a series of tubes (blood vessels). The flow is maintained by a central pump, the heart, which really is an enlarged piece of tubing with strong, thick, muscular walls. The heart pumps blood into arteries and these supply the organs. Blood drains away from the organs and returns to the heart along veins. Between each artery and vein, in every organ, blood is channelled along a network of blood capillaries. Blood cannot flow backwards because of one-way valves in the veins and in the heart itself.

The heart is actually a double tube, and controls two circulations – the blood supply to and from the lungs (pulmonary circulation) is separate from that supplying all other parts of the body.

The heart pumps blood into the arteries in a succession of pressure waves towards the organs. The

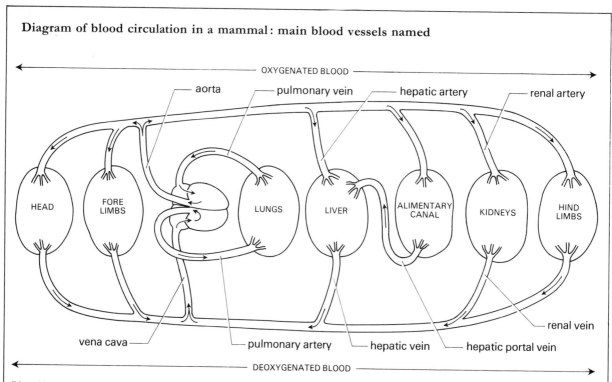

Diagram of blood circulation in a mammal: main blood vessels named

Blood leaves heart along arteries; blood returns to heart along veins. Capillary networks inside organs between arteries and veins. Blood passes through the heart *twice* during each circulation because pulmonary circulation (the lungs) is separate from the rest.

Principles of Biology

capillary networks in the organs are very narrow and form bottle-necks, rather like the turnstiles at football grounds. The narrowness of the capillaries creates friction, slowing the blood flow. The blood pressure drops and is very low by the time the blood reaches the vein taking blood away from the organ.

Capillaries pass close to the living cells of the body; in fact, the living cells are separated from the blood in the capillary by small pools of liquid (called tissue fluid) and by the thin walls of the capillaries themselves. These thin walls are 'leaky' and allow the capillaries to pass essential materials from the blood to the cells around them – oxygen diffuses out of the red corpuscles; water with dissolved salts, glucose, amino-acids, and other substances, diffuses from the plasma into the tissue fluid and then into the cells. Even some white corpuscles can squeeze out between the cells of the capillary walls to deal with disease in the body. Carbon dioxide and other waste substances produced by the living cells diffuse into the tissue fluid outside.

Some of the fluid lost from the blood is picked up by the capillaries from the tissue fluid spaces towards the 'vein-end' of the capillaries. Most of the lost fluid is collected by small tubes which belong to the lymph system, a system of tubes closely associated with the blood. This lymph fluid is eventually emptied into large veins just above the heart. In this way, the blood volume is kept quite constant.

Blood groups
Blood contains its own proteins. The different blood groups to which human beings belong are due to differences in some of these blood proteins. The four main groupings are A, B, AB and O. They are inherited. After an accident or an operation in hospital, the patient's blood group must be known, otherwise a transfusion could cause death. A patient with blood group AB can safely receive blood from any other group. Blood of group O is very useful in a hospital blood-bank because it can safely be given to anyone. In all other cases, different bloods mixed together will react, causing the red corpuscles to stick together and block the small blood vessels. The patient's blood can recognize the blood proteins of the different group. (Transplant operations may similarly fail because the protein in the patient's tissues recognizes, reacts with and rejects the 'foreign' protein in the transplanted organ.)

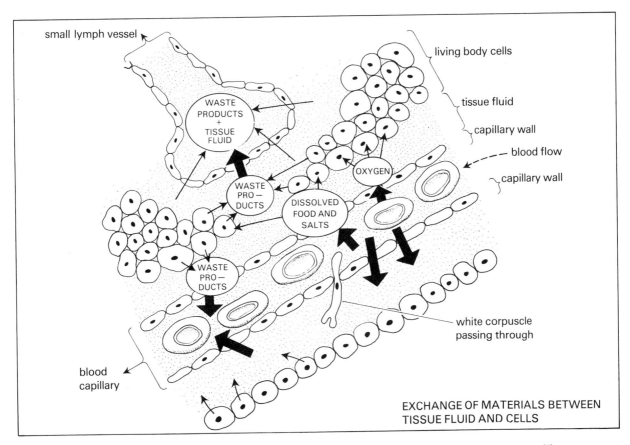

EXCHANGE OF MATERIALS BETWEEN TISSUE FLUID AND CELLS

Three main types of blood vessel

Arteries

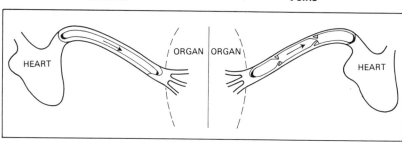

Blood carried *away from heart* to organs

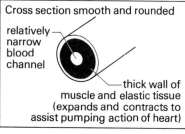

Blood flows fast, in pulses, under high pressure

Blood is oxygenated (except in pulmonary artery)

Situated deeply

Veins

Blood carried *back to heart* from organs

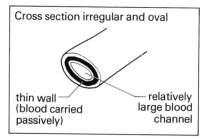

Blood flows quite slowly, smoothly, under low pressure

Blood is deoxygenated (except in pulmonary vein)

Situated nearer body surface

Capillaries

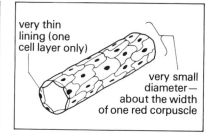

Link arteries with veins in the tissues and organs

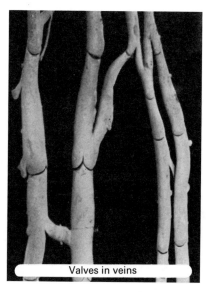

Valves in veins

Valves, like pockets on the inner walls, fill to allow movement of blood only towards the heart

Blood seeping backwards flows into the pockets — the pockets become larger and block the vein

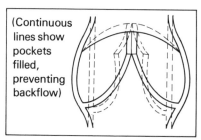

(Continuous lines show pockets filled, preventing backflow)

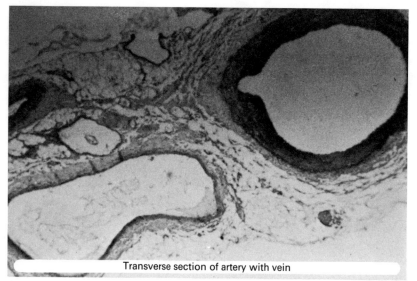

Transverse section of artery with vein

86 **Principles of Biology**

The composition of blood
Blood has cell-like structures (corpuscles), suspended in a fluid.

Plasma

A pale yellow liquid, mainly water with dissolved foods (glucose, amino-acids, fatty substances), mineral salts, hormones, urea, special blood proteins, traces of oxygen and carbon dioxide

Platelets

In man, 1 mm³ contains 250 000

Manufactured in bone marrow

Sticky fragments of protoplasm

Granules, but no nucleus

Assist in clotting

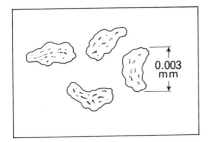

White corpuscles

In man, 1 mm³ contains 10 000 (ratio of reds to whites = 500:1)
Several different types, some formed in bone marrow, others in lymph glands

Nuclei of different shapes

Most can move (like an amoeba) independently of the blood flow

Concerned with defence against disease

Red corpuscles

In man, 1 mm³ contains 5 million

Manufactured in bone marrow – about 1 million/second

Life span about 12–15 weeks, then destroyed in liver and spleen

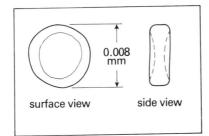

Disc-shaped, curved inwards on both surfaces

No nucleus

Filled with iron compound, haemoglobin, in a sticky elastic envelope (haemoglobin carries oxygen as oxy-haemoglobin: when oxygen is given up, haemoglobin is able to return to lungs and pick up more)

Often piled together:

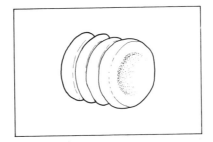

The composition of lymph
Lymph is essentially the same as blood, but without red corpuscles – red corpuscles cannot pass out from capillaries.

Blood as a means of transport and defence

Defence against disease

Bacteria and other 'foreign' particles that enter the blood are dealt with. Many *white corpuscles* are attracted to these intruders, and to any dead cells, and flow towards them. The corpuscles ingest them (take them in) and break them down in the same way that *Amoeba* feeds.

When part of the body becomes infected, its blood supply increases. More warm blood leads to a local rise in temperature. This is the cause of *inflammation*. The higher temperature, by itself, may kill many bacteria. At the same time, the capillary walls allow white corpuscles to enter and leave the blood more easily.

Blood (and lymph) contains two substances – *antibodies* and *antitoxins* – which help the white corpuscles fight disease. Some antibodies stick bacteria together, making them immobile and unable to enter the tissues. Other antibodies make bacteria easier to ingest or even attack bacteria and digest them. Different diseases are fought by different antibodies. Antitoxins are chemicals that react with the poisons (toxins) excreted by many bacteria. Toxins are often the cause of disease symptoms – rash, fever, etc. – but antitoxins make them harmless. White corpuscles may be killed by toxins – pus contains dead corpuscles as well as dead body cells and bacteria.

Clotting reduces bacterial invasion after a wound. It begins when

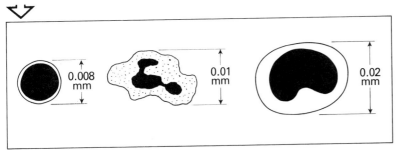

Transport 87

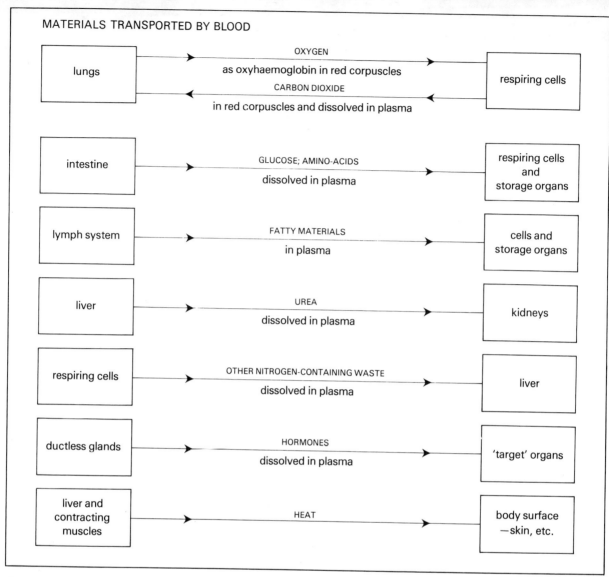

damaged platelets and other cells nearby break down and release chemicals that trigger off a train of events involving special plasma proteins. Eventually a dissolved protein called fibrinogen is changed to fine threads of fibrin. This forms a mesh in which blood cells are trapped, dry and harden into a scab. New growth of tissue under the scab seals the wound.

With many diseases, the ability to manufacture antibodies remains in the body long after recovery. This means that the same disease is seldom suffered twice, or that the second attack is much less severe than the first – the body develops *immunity*. This knowledge is used in vaccination and inoculation, where a mild form of the disease is given to stimulate antibody production. This is artificial immunity.

Functions of lymph

Lymph has the same functions as blood except that it has no red corpuscles to transport oxygen. Lymph acts as a link between blood and the living cells, supplying them with essential materials and removing their waste.

The heart and the circulation of blood

Contractions of the heart drive blood into the arteries. When the heart muscles relax between contractions, blood is drawn in from the large veins and backflow is prevented by sets of valves inside the heart.

Pressure drops as blood flows along capillaries, so that there is not enough force in the veins to return blood to the heart against gravity. During breathing in, pressure drops in the chest cavity and this helps to draw blood into the heart. But the main factor causing

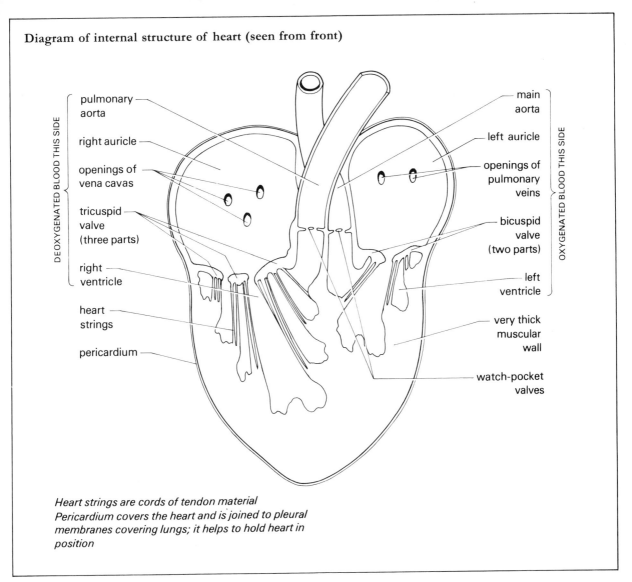

Diagram of internal structure of heart (seen from front)

DEOXYGENATED BLOOD THIS SIDE

- pulmonary aorta
- right auricle
- openings of vena cavas
- tricuspid valve (three parts)
- right ventricle
- heart strings
- pericardium

OXYGENATED BLOOD THIS SIDE

- main aorta
- left auricle
- openings of pulmonary veins
- bicuspid valve (two parts)
- left ventricle
- very thick muscular wall
- watch-pocket valves

Heart strings are cords of tendon material
Pericardium covers the heart and is joined to pleural membranes covering lungs; it helps to hold heart in position

the return of blood to the heart is the contraction of muscles moving the skeleton. All the large veins pass close to these muscles which, of course, thicken when they contract: this presses on the veins. Blood cannot flow away from the heart because the veins are closed as their pocket-valves fill.

The heart is really a pear-shaped hollow muscle, placed slightly towards the left, between the lungs. Two chambers on the right contain deoxygenated blood and two completely separate chambers on the left contain oxygenated blood.

Blood enters the two upper chambers (auricles) and leaves from the two lower (ventricles). Four chambers are needed because of the separate blood circulation to the lungs. Blood from the lungs enters the left auricle (along pulmonary veins). All other blood enters the right auricle (along the vena cavas). Each auricle then feeds blood into a ventricle. Blood leaves the right ventricle for the lungs (along the pulmonary aorta) and leaves the left ventricle for elsewhere in the body (along the main aorta).

The two sides of the heart function together. When the auricles contract, the ventricles relax; blood is pumped from auricles to ventricles. When the ventricles contract, blood is pumped into the the aortas; at the same time, the auricles relax and more blood enters.

Valves close during contraction of the ventricles, so that blood cannot return to the auricles. The tricuspid (right side) and bicuspid (left side) valves are like parachutes, prevented from turning inside-out under great pressure by heartstrings. Three-part watch-pocket valves, at the lower end of

Transport

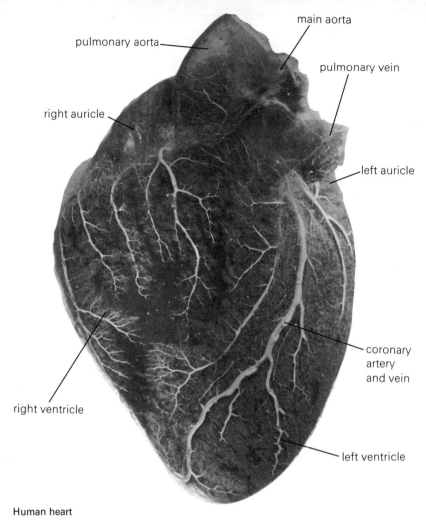

Human heart

depends on age and activity. Heartbeat rate is increased by emotional excitement (the heart is directly affected by the hormone adrenalin), shock, fever, very warm weather and certain drugs.

Heartbeat varies with age (man)
At birth   140 beats/min
At puberty   80 beats/min
In old age   60 beats/min

Heart performance varies with circumstance (man)

	Beats/min	Output/min
At rest	70	5 litres
During strenuous exercise	150	23 litres

Transport in flowering plants

Water is

- absorbed from the soil by the roots;
- lifted up the stem;
- evaporated by the sun's energy from the leaves.

Root-hairs, special cells in the outer layer of the root, absorb water. Each cell forms a long projection outwards from the root and grows between the soil particles: this provides a large surface area for more efficient absorption.

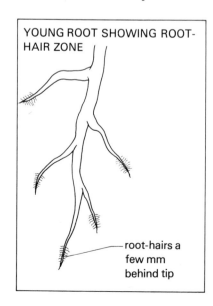

YOUNG ROOT SHOWING ROOT-HAIR ZONE

root-hairs a few mm behind tip

each aorta, close to prevent blood seeping back into the ventricles each time the ventricles relax. These valves have no strings because the pressures here are much less. The ventricle walls, particularly on the left, are much more muscular than auricle walls.

The sounds of heartbeats are produced by the valves closing. The loud, dull thud occurs when the cuspid valves close and tension on the heartstrings increases. The shorter, sharp, rather flapping sound that follows is caused by the watch-pocket valves closing. On average, the heartbeat cycle is completed in 0.8 secs, so that there are 75 beats/min. The heart muscle relaxes for 0.4 secs between each beat.

Beating begins before birth and may continue for some hours after death. Heart muscles contract without nervous stimulation; instead, they are stimulated by a 'pacemaker' in the right auricle wall – this can be replaced by an artificial electrical pacemaker if the heart becomes defective. The *rate* of heartbeat, though, is controlled by the nervous system, acting through the pacemaker.

The performance of the heart

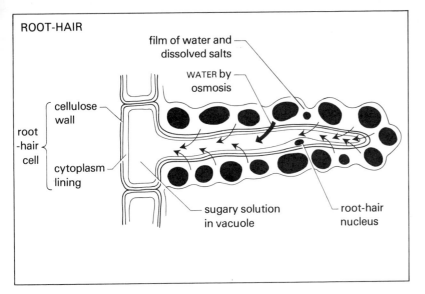
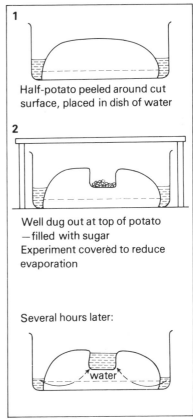

Root-hairs live for no more than a few days and are constantly replaced as the root grows.

Water is absorbed by *osmosis*. A dilute solution of mineral salts forms a thin film around the soil particles. The sap in the root-hair provides a more concentrated sugary solution, and the root-hair cytoplasm forms a semipermeable membrane.

Although root-hairs may take in salts passively by diffusion, it is known that roots have to work in order to obtain enough mineral salts. The energy for this comes from respiration – plants grown in solutions of mineral salts will not thrive unless the solutions are oxygenated.

The absorbed water and dissolved salts are then carried up the plant in the xylem tissue of the veins. This can be shown by carefully washing soil away from the roots of a healthy plant and placing the roots in coloured water for several hours. Cross sectioning shows that the stem becomes stained only in the position of the xylem.

Plants have no internal pump such as a heart. Because the xylem consists of narrow tubes, fluids are lifted a few metres by capillary action. But the main energy source to lift fluid against gravity (to a height of many metres in a tall tree) comes from outside, from sunlight. It is the result of plants losing water vapour into the surrounding air by a process called *transpiration*.

Most plants lose little water when the air is cool, but evaporation becomes rapid when the sun warms the air around them. (Rather like sweating, this cools the plant.) The vapour diffuses out mainly through the leaf pores (stomata) but some is also lost through similar pores in the stem.

The xylem tubes pass up the stem and branch into the leaves. Here, they are separated from the surface pores by many living cells. There is also an air space just inside each pore. As water evaporates through the pore, water is drawn across the leaf cells from the xylem tubes, to take its place. This is done by osmosis: the experiment at the top, right, shows how.

After several hours, the sugar forms a solution that rises in the well. The water level in the dish falls correspondingly. Water seems to have passed upwards across the potato cells from dish to well.

The explanation can be simplified by imagining that the potato has only three cells in a line between dish and well (see below).

The sugar dissolves in the sap of the cells damaged when the well is dug. This forms a concentrated solution, drawing water out of cell A by osmosis. As this cell

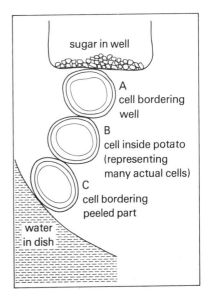

Transport 91

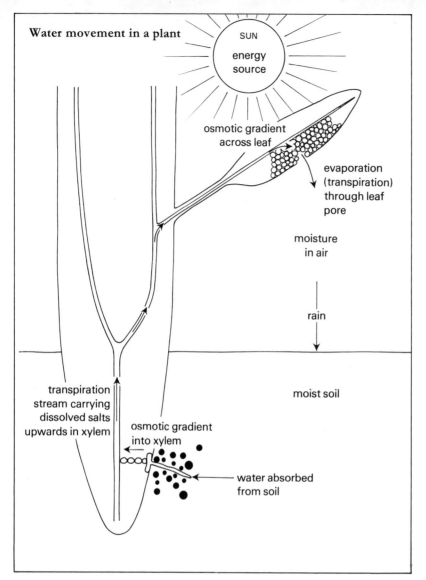

loses water, *its* sap becomes more concentrated and draws water out of cell B. Cell B now has more concentrated sap and draws water out of cell C. Water in the dish passes into cell C to replace that lost to the other cells. In this way, water passes from dish to well. An *osmotic gradient* is set up across the cells.

If the experiment is carried out with a boiled potato, osmosis does not occur – the cells have been killed and their semipermeable membranes destroyed.

The leaf cells lose water by exactly the same method that the root cells gain it – by osmosis. In this case, because the leaf cells are evaporating water to the outside air (transpiration), the osmotic gradient operates in the opposite direction. Water, carrying dissolved salts, is drawn upwards in the xylem to replace that being lost. This upward movement is the *transpiration stream*.

The xylem tubes at the root centre are separated from the root-hairs on the outside by large numbers of living cells. Another osmotic gradient carries absorbed water to the xylem, where it joins the upward transpiration stream.

In strong sunshine, evaporation occurs faster and water moves more quickly between the cells. The transpiration stream moves faster and more water is absorbed by the roots if water is available in the soil. A large tree may transpire several hundred litres of water during one summer's day. In most plants, the pores close at night, and transpiration slows.

Transpiration increases not only when the sun is brighter and warmer. A strong wind blowing across the leaf surface blows away the water vapour, so that more can evaporate. But when the air around the leaf is very humid (laden with water vapour), as after heavy rain, transpiration is slow. Young trees and shrubs may turn brown and die because of water-loss during dry, windy periods.

Some plants can survive where water is scarce, by reducing transpiration. Water is not available to plants in deserts and in cold regions either in land well away from the equator or at high altitudes, because plants cannot absorb from the soil when the temperature is low. (They may still lose water by evaporation – food dries even in a refrigerator.) Pine trees have needle-like leaves with very little surface for transpiration. The leaves of grasses on sand dunes are rolled inwards.

Leaf-fall
Evergreens such as holly have relatively few pores and thick waxy cuticles; this restricts water loss. Their leaves are shed at any time of year. Deciduous plants – horse-chestnut, ash, hawthorn – shed their leaves each autumn when the plant becomes dormant. This prevents too much water loss through the pores and thin cuticle during sunny but frosty weather in winter. It provides a method of excreting

Experiment to find out whether roots absorb water

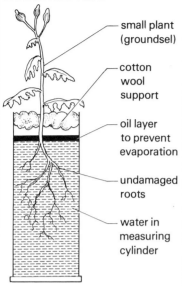

- small plant (groundsel)
- cotton wool support
- oil layer to prevent evaporation
- undamaged roots
- water in measuring cylinder

After several hours in a warm atmosphere, the water level has fallen.
Control experiment—set up at the same time, exactly similar but with no plant: no change in water level.

Experiment to find out if plants give off water vapour

- large plastic bag, completely covering but not touching the potted plant
- plastic cover for soil and pot—prevents water loss from soil
- Control: similar but no plant

After an hour in warm conditions, droplets of colourless liquid form on inside of bag. Careful removal of bag allows liquid to be tested: white anhydrous copper sulphate is added. A colour change to blue shows that the liquid is water. No liquid appears in the control.

Experiment to find which leaf surface loses more water

A series of fresh leaves, their surfaces smeared with petroleum jelly, are tied by their stalks to a 'clothes line' thread suspended between two retort stands.

Result after two hours in a warm atmosphere:

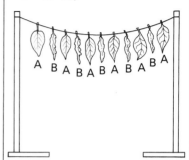

A—smeared on lower surface only
B—smeared on upper surface only
The leaves with their upper surfaces smeared shrivel and dry more quickly—most leaves have more stomata below than above.

Experiment to find out whether roots absorb dissolved substances

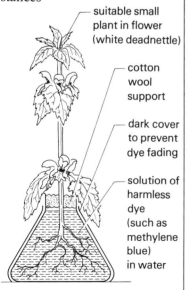

- suitable small plant in flower (white deadnettle)
- cotton wool support
- dark cover to prevent dye fading
- solution of harmless dye (such as methylene blue) in water

After several days, blue lines appear in the petals—the dye has passed up the stem from the roots into the veins of the petals.

A potometer

This apparatus measures the rate of water absorption in a cut leafy shoot. Water absorption in this case is equivalent to water loss by transpiration.

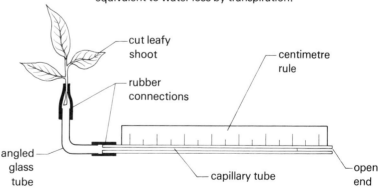

- cut leafy shoot
- centimetre rule
- rubber connections
- angled glass tube
- capillary tube
- open end

The glass tubes are filled with water in a sink. A small leafy shoot is cut from a larger shoot under water in the sink to prevent air entering the xylem. The small shoot is fixed in position, the joint made airtight, and the apparatus taken from the sink. As the shoot absorbs water, air enters the capillary tube. If the tube's diameter is known, the volume of water absorbed can be estimated from the rate at which the water retreats along it. Comparisons can be made by placing the apparatus in different positions—bright and dim light, still and moving air, etc. Each time, the apparatus must be returned to the sink, the capillary tube disconnected and refilled with water.

Transport

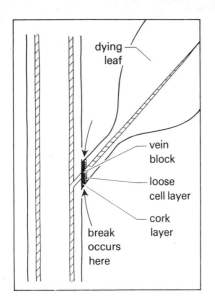

The veins of plants contain two different types of tube. The xylem tubes carry water and salts upwards. The phloem transports food made in the leaves to all other parts – stems, roots, flowers, fruits. This process is called translocation.

accumulated waste from the summer's growth. Just before leaf-fall, materials are withdrawn from the leaf and waste materials passed into it. The veins leading into the leaf are sealed and the leaf dies. A layer of loose, fragile cells forms at the leaf joint.

Below this, a thin layer of insulating cork forms. The cells in the loose layer gradually break down as the wind jostles the leaf and tears it away from the branch. The cork layer seals the wound and leaves a scar. This prevents water loss and entry of disease.

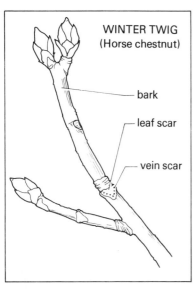

Energy production

Living organisms produce energy by the process called respiration. Like machines, living organisms are really energy converters: in this case the fuel is food. When food is broken down, energy is set free, to be used for movement, the manufacture of more protoplasm in

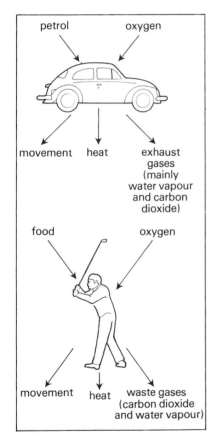

growth, the secretion of essential substances or sending messages along nerves.

Some of the energy is released as heat. In birds and mammals this helps to keep the body temperature constant. Respiration provides a useful test for the peculiar processes which we call 'life': anything that is respiring is alive.

In most organisms respiration requires a supply of oxygen gas – *aerobic respiration*. Sometimes chemical compounds containing oxygen can be broken down without using oxygen gas – *anaerobic respiration*.

During aerobic respiration, land-living organisms must obtain oxygen from the air around them; aquatic life has to use oxygen dissolved in the surrounding water. Various mechanical processes have developed (ventilation of the lungs in mammals and of the gills in fish) to collect the necessary oxygen. These processes are also used to remove the waste products which respiration produces. All these processes are *breathing*.

True respiration is a chemical process taking place inside all living cells – it is sometimes called *internal* or *cellular respiration* to distinguish it from breathing. It is complex, but the equation shows what is used up and what is produced during aerobic respiration:

simple sugars	$C_6H_{12}O_6$
burnt with	+
oxygen	$6O_2$
produce	↓
carbon dioxide	$6CO_2$
and	+
water vapour	$6H_2O$
and	+
energy	ENERGY

Because the sugars are broken down and used up, this means that respiration involves a loss of weight.

Several aspects of the equation can be tested by experiments (p. 96).

Anaerobic respiration takes place where oxygen gas is not available or is in short supply. It is a relatively wasteful process and produces much less energy. Muscles are capable of anaerobic respiration. For example, during a strenuous race, the blood cannot carry oxygen to the muscles fast enough and glucose is broken down anaerobically to lactic acid. The build-up of this waste product results in some of the unpleasant effects of vigorous exercise – for example, being out of breath. The muscles gradually build up an 'oxygen debt' which has to be paid off by continuous rapid deep breathing afterwards – the lactic acid is oxidised to carbon dioxide and water:

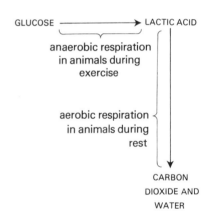

Anaerobic respiration in plants produces carbon dioxide and alcohol as waste products. Seeds will respire anaerobically for a short time if deprived of oxygen. Under these conditions, though, they will never germinate and eventually die, poisoned by their own waste products:

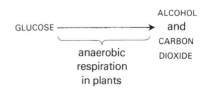

Experiment to find out whether respiration produces carbon dioxide

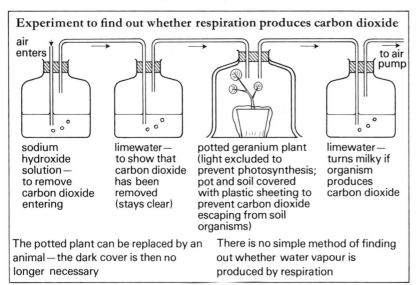

- **sodium hydroxide solution** — to remove carbon dioxide entering
- **limewater** — to show that carbon dioxide has been removed (stays clear)
- **potted geranium plant** (light excluded to prevent photosynthesis; pot and soil covered with plastic sheeting to prevent carbon dioxide escaping from soil organisms)
- **limewater** — turns milky if organism produces carbon dioxide

The potted plant can be replaced by an animal — the dark cover is then no longer necessary

There is no simple method of finding out whether water vapour is produced by respiration

Experiment to find out whether respiration produces energy

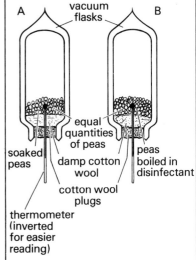

- vacuum flasks
- equal quantities of peas
- soaked peas
- damp cotton wool
- peas boiled in disinfectant
- cotton wool plugs
- thermometer (inverted for easier reading)

Because the peas in B are dead, the thermometer will record only changes in air temperature. Any larger temperature change in A will be due to heat energy produced by the respiring peas.

Experiment to find out whether oxygen is taken up during respiration

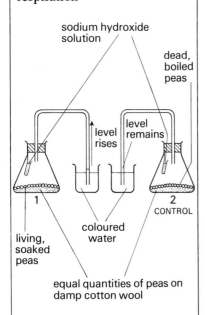

- sodium hydroxide solution
- dead, boiled peas
- level rises
- level remains
- coloured water
- 1 living, soaked peas
- 2 CONTROL
- equal quantities of peas on damp cotton wool

The sodium hydroxide in the ignition tubes absorbs the carbon dioxide given out by the respiring peas in 1. The level of coloured liquid rises considerably in the side tube as the peas remove a gas from the air. After several days, a lighted splint put into flask 1 is extinguished, showing that the air in here has much less oxygen. There is little change in the control flask and a splint will remain alight in it.

Experiment to find out whether there is a loss of dry weight during respiration

Weigh 100 dry peas and divide weight by 100 to find average weight of pea. Soak all the peas in water for 24 hours, then kill 50 peas by boiling. Place the two groups (boiled and unboiled) separately in a warm atmosphere on damp cotton wool. At regular intervals (every second day), remove 10 peas from each group and dry overnight in an oven kept thermostatically at 100°C. Each group of 10 peas is then weighed.

It should be found that the weight of the boiled peas remains constant, but that of the unboiled peas decreases steadily from the start of germination.

Experiment to demonstrate fermentation

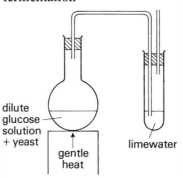

- dilute glucose solution + yeast
- gentle heat
- limewater

The mixture in the flask gives off a colourless gas which turns the limewater milky.

Experiment to find out whether peas will respire without oxygen

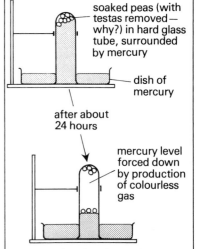

- soaked peas (with testas removed — why?) in hard glass tube, surrounded by mercury
- dish of mercury
- after about 24 hours
- mercury level forced down by production of colourless gas

If pellets of sodium hydroxide are placed into the tube, the gas is absorbed and the level rises — carbon dioxide has been produced without the presence of oxygen.

Control experiment consists of similar apparatus at same time, but with dry peas — the mercury level remains unchanged.

For certain simple organisms, the normal method of respiration is anaerobic. In yeast and some bacteria the process is important commercially and is called *fermentation*. Yeast can be used to produce wines because it does not kill itself until the alcohol it has produced approaches fifteen per cent. (Most living cells are killed by much less alcohol than this.) This is sufficient to make table wines, but to produce spirits (whisky, gin, vodka, etc.) the alcohol is distilled to provide a higher alcohol content. Fermentation by yeast is also used to make bread: the yeast ferments the dough, setting free bubbles of carbon dioxide which become trapped in the bread and make it rise.

Breathing

BREATHING (EXTERNAL RESPIRATION) *is the process of obtaining the oxygen needed for aerobic respiration and getting rid of waste products such as carbon dioxide.* Oxygen enters all living cells by diffusion, in solution, in water. It enters faster at higher temperatures and also when there is more oxygen outside the cell. The amount needed depends on the size of the cell, but the amount that can enter depends on the total surface area of the cell. A single cell can take in oxygen on all sides. Most of the many cells in a large organism are joined to other cells, so that there is much less surface available for oxygen entry. Also, these cells are not in contact with the outside air or water at all.

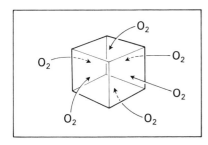

Because of this, large animals require a special respiratory surface in contact with the outside, and large enough to absorb sufficient oxygen for all the body's cells. In man and other mammals, this is provided by the lungs. There must also be a blood system to carry the absorbed oxygen rapidly to every cell (see below).

In mammals, the lungs are connected with the atmosphere by a sequence of pipes to form a respiratory system. Air entering the nose is warmed, and small hairs and a continually moving film of mucus trap unwanted particles. The filtered air then passes down the throat and into the windpipe (trachea). The windpipe is enlarged at its upper end, forming the voice-box (larynx), which has vocal cords stretched across. These produce sounds when forced to vibrate. The windpipe extends down the neck into the chest cavity. It is kept open, even when air is being drawn in (or when food is passing down the gullet immediately behind the windpipe) by rings of cartilage which are incomplete on the side next to the gullet. At its lower end, the windpipe forks into two bronchi, one entering each lung.

Inside a lung, the bronchus subdivides many times forming bronchioles, each of which ends in a cluster of 'bubbles', the air-sacs (alveoli). A human being is said to contain seven million alveoli – the lungs are spongy and elastic, honeycombed with air passages and spaces. The lungs also have a very good supply of blood capillaries: the pulmonary arteries enter the lungs and the pulmonary veins leave.

The lungs are bounded by the curved ribs in front and at the sides, and by the diaphragm below. The ribs are held together by muscles, forming a cage. The inner surfaces of the ribs and diaphragm are lined

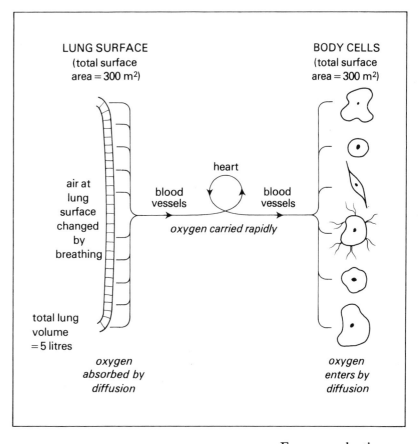

Energy production

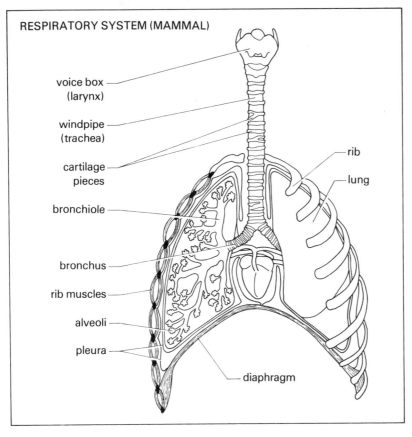

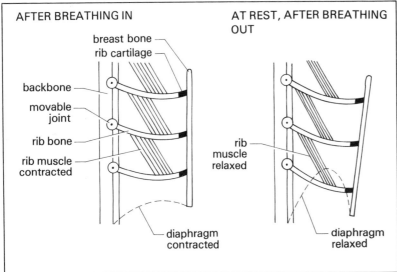

	Air breathed in	Air breathed out
Nitrogen	77%	no change — 77%
Oxygen	21%	13%
Water vapour	1.5%	6%
Carbon dioxide	0.004%	4%

by a sheet of tissue; this sheet also doubles back over the outer surface of the lungs. The two sheets form the pleura and the space between is filled with pleural fluid, so that the sheets cannot easily separate.

The system works rather like bellows. The ribs are able to swivel where they are jointed to the backbone. When the rib muscles contract, they pull on the ribs and the whole rib cage (with the breastbone in front) moves upwards and outwards. At the same time the diaphragm muscles contract. At rest, the diaphragm is a dome-shaped sheet of tissue, but it flattens towards the abdomen when it contracts. The actions of the ribs and diaphragm increase the volume of the chest cavity. The lungs are forced to expand because their outer surface is attached to the rib cage and diaphragm by the pleural membranes. Air then flows into lungs to fill the extra space.

Breathing out occurs when these actions are reversed. The muscles relax: the rib cage swings down and in, and the diaphragm dome collapses into the chest cavity. The lungs are compressed and air is forced out.

In the alveoli, gases are exchanged between the air and the blood. The air in the alveoli is separated from the flow of blood in the lung capillaries by only two thin cell layers – the walls of the alveoli and the walls of the capillaries. Oxygen dissolves in the moist inner lining of the alveoli before diffusing through to the red corpuscles. Carbon dioxide diffuses in the reverse direction. (See top, p. 99.)

The chart shows how much the composition of air is altered in the lungs of man. With quiet breathing, only half a litre (approximately) of air passes into and out of the lungs at each breath – this is 'tidal air'. The lungs remain half-full (two-and-a-half litres) between breaths: the air is not completely changed. Forced breathing can empty the lungs more completely, leaving only about one-and-a-half litres. Fully-trained athletes can achieve this and can breathe in up to three-and-a-half litres of air – this is the maximum 'vital capacity' of the lungs.

Normally, a man takes about fifteen breaths a minute. The rate is controlled by various reflexes, though the rate can, of course, be altered deliberately; for example, when shouting. The brain is very sensitive to small changes in carbon dioxide concentration in the blood. When carbon dioxide accumulates in the blood (for example after strenuous exercise), the brain triggers off faster and deeper breathing movements.

In plants, there is no special breathing mechanism nor breathing movements like those found in most animals. Plants are much less active; they do not move from place to place and they produce relatively small amounts of energy. No cell in a plant is very far from the outside air and all the necessary oxygen can be obtained by simple diffusion. Leaves are thin and flat. They have large surface areas and small volumes, and are ideal for the exchange of gases. The surfaces are riddled with small pores (called stomata) which lead into a honeycomb of air passageways between the actual leaf cells.

Soft green stems also have stomata, but in trees and shrubs the stomata are replaced by larger craters called lenticels. With the unaided eye, they can be seen as small 'spots'. Again, there is a system of passageways leading inwards.

Living plant cells are not watertight: they continually leak. The oxygen in the air passages dissolves in the cell's surface moisture and passes into the cell in solution (as in the air-sacs of the lungs). More

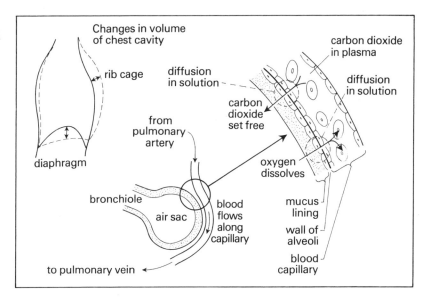

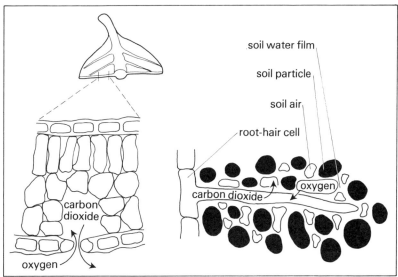

can diffuse in as soon as the oxygen is used up. At the same time, carbon dioxide diffuses from the cells and out into the air. During the hours of sunlight, photosynthesis takes place and the gases reverse their direction – carbon dioxide diffuses in and oxygen diffuses out.

Roots obtain oxygen from the soil air. This dissolves in the soil water and diffuses into the root-hairs; the water that passes in by osmosis also contains dissolved oxygen. Carbon dioxide diffuses out into the soil air. In addition, the air passageways of the stem connect with similar passageways in the root.

Part of a woody stem (in section) showing a lenticel

Energy production

Removal of waste

Living organisms are converters of energy (so are motor vehicles): they use food as fuel, as an energy source. The energy is released during respiration and waste materials are produced.

General 'wear and tear' produces waste, in living organisms as well as in machines. Mechanical parts have a limited life – they eventually wear out, are thrown away, and replaced. The red blood cells in a mammal also have a limited life. After several weeks they are destroyed and replaced. Some of their component parts are 'thrown away' by the body (though much of the material is recycled). All these waste materials are removed by a process called excretion.

EXCRETION is the removal of waste formed by chemical processes inside living cells.

The removal of undigested or unabsorbed food is *not* excretion. These materials have not at any time entered the cells or taken part in the cells' living processes. They are removed by egestion.

Animals cannot re-use the materials produced by respiration, except perhaps water. In the green parts of plants, though, these materials can become the raw materials of photosynthesis.

When animals eat protein food, it is digested into a range of amino-acids, some of which are not needed by the animal for making its own protein. This surplus is converted to other uses in the liver, at the same time forming waste materials which have to be excreted. Green plants make the proteins they need from simple chemicals.

Because of these differences, excretion is in some ways a simpler process in green plants.

The lungs, skin and kidneys are the three main excretory organs of mammals, though the intestine carries out some excretion as well as egestion. All these organs, except the kidneys, are described elsewhere in this book.

The kidneys

The kidneys act as blood filters. In addition to removing such waste as urea from the blood, they control the water and salt content of the blood – they are *osmoregulators*.

Blood enters each kidney along a renal artery (a branch of the main aorta). It leaves by a renal vein, which feeds into the vena cava. Inside the kidney, the artery divides into a series of small blood vessels which make contact with about one million minute kidney tubules whose walls are one cell thick. The filtering takes place here, before the small vessels join and form the renal vein.

Internally, the outer kidney region (cortex) is quite distinct from the inner medulla because the

MAIN FORMS OF EXCRETION		
Green plants		
Material excreted	*Source of material*	*Method of excretion*
carbon dioxide	respiration	gases diffuse out from the cells — carbon dioxide at night, oxygen during the day — pass along the air spaces between the cells and out through the pores of the stem and leaf
oxygen	photosynthesis	
mineral salts	regulation of composition of cell fluids	stored as insoluble crystals or in the dead (heart) wood of trees OR passed into leaves and lost at leaf-fall
Mammals		
Material excreted	*Source of material*	*Method of excretion*
carbon dioxide	respiration	carried by blood to lungs, and exhaled
urea	breakdown of excess amino-acids in liver	carried by blood to kidneys (and skin), and removed
bile pigments	breakdown of dead red corpuscles in liver	carried by bile into intestine and removed with faeces
mineral salts	regulation of composition of body and cell fluids	carried by blood to kidneys (and skin), and removed

Principles of Biology

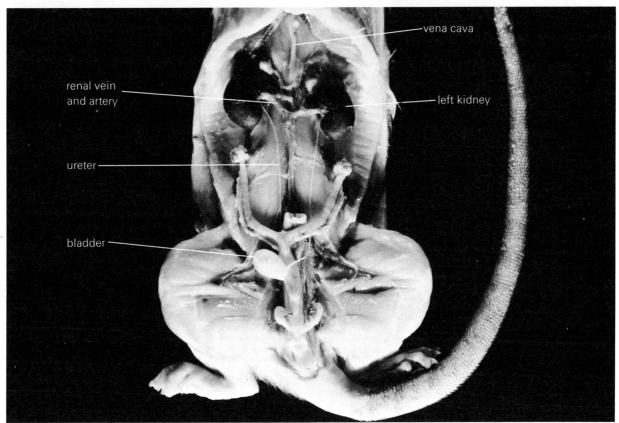

Kidneys are dark red, compact organs, well behind most other abdomen organs, one on each side of backbone: usually covered by a protective cushion of fat.

kidney tubules have different regions – the straight parts of the tubules in the medulla are absent in the cortex.

Materials filtered from the blood leave the kidney along the ureter. This passes out of the kidney at the curved inner surface, near where the renal artery and vein enter and leave. The ureter begins inside the kidney as a small space (pelvis) before it narrows into a fine white tube leading to the bladder.

A branch of the renal artery supplies each kidney tubule. It passes into the bowl of Bowman's capsule, forming a tight knot of looped capillaries (a glomerulus). The blood vessel which leaves is narrower than the one which enters; the resulting bottleneck in

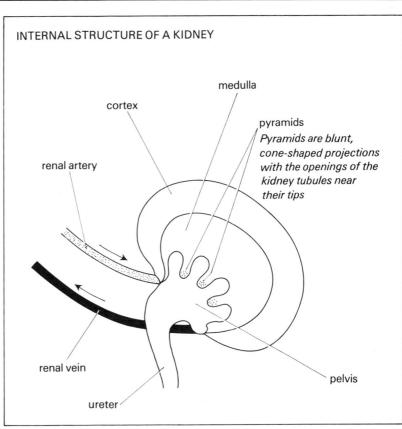

INTERNAL STRUCTURE OF A KIDNEY

Pyramids are blunt, cone-shaped projections with the openings of the kidney tubules near their tips

Removal of waste

the capillary loops maintains a high blood pressure. Small molecules are forced (by heart and arterial pressure) from the bloodstream into the kidney tubule; this filtering under pressure takes place through the thin walls of the capillaries and of Bowman's capsule. The small blood vessel leaving Bowman's capsule then passes close to the other parts of the tubule before becoming part of the renal vein. The diagram shows that, besides urea, the filtered solution contains valuable materials. As it flows down the tubule its composition alters. The respiring cells lining the tubule use energy to absorb certain materials and pass these back into the blood capillaries close to them:

1 **Water** – the solution is gradually concentrated (about 60 times) as it flows between Bowman's capsule and collecting duct. It is known that reabsorption in the loop region depends on the amount of water in the body. The brain is sensitive to the water content of the blood. When large quantities of fluid are drunk, the blood becomes diluted, and instructions are sent to the kidneys so that very little water is reabsorbed – this results in the production of a large volume of dilute urine for several hours. Kidney output depends on water loss from other parts of the body, particularly the sweat glands. In hot weather or after exercise, the sweat glands remove more water from the blood. The kidney reabsorbs much more water – this results in the production of only small quantities of concentrated urine (it is accompanied by the sensation of thirst). The kidney is performing its important function of osmoregulation.

2 **Dissolved substances** – the coiled tubules reabsorb glucose, amino-acids and some salts, in addition to water. Normally, urine does not contain glucose or amino-

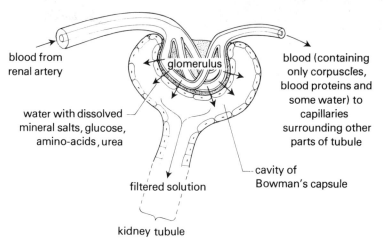

DIAGRAM SHOWING FILTERING INTO BOWMAN'S CAPSULE

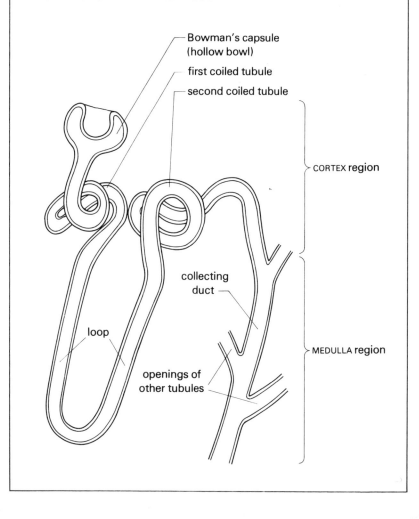

DIAGRAM OF SINGLE KIDNEY TUBULE

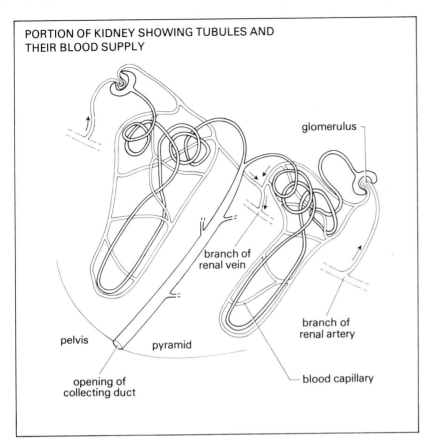

PORTION OF KIDNEY SHOWING TUBULES AND THEIR BLOOD SUPPLY

acids, though there are traces of some salts such as sodium chloride. However, in diseases such as diabetes, the process of reabsorption fails to deal with the large amounts of glucose reaching the kidney tubules, so that glucose is found in the urine. Doctors diagnose diabetes by testing a urine sample for glucose.

The liquid in the collecting duct is very different from that which filtered into Bowman's capsule. Urine, a slightly acid liquid, leaves the collecting ducts at the tips of the pyramids and passes into the pelvis. It is forced down the ureter by waves of peristalsis and is continually squirted into the bladder (a man can store nearly half a litre).

As the bladder fills, its walls gradually extend. Eventually, the ring of muscle (sphincter) controlling the exit relaxes and muscle contraction forces urine along the urethra and out of the body. Control of these muscles, which allow the release of urine, has to be learnt. On average, a man produces about a litre-and-a-half of urine per day.

Removal of waste

The skin

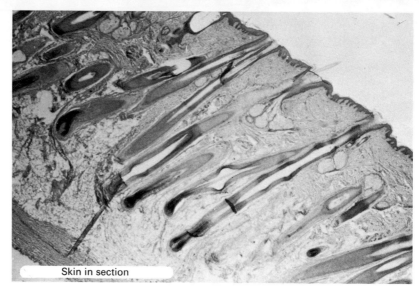
Skin in section

The skin is a boundary layer separating the internal world of the body from the continuously changing world outside. It has several important features:

- It is elastic, flexible and waterproof (like a frogman's suit)
- It prevents loss of body heat (like an overcoat)
- It can get rid of body heat, even when the outside temperature is higher than the body's temperature (like a refrigerator)
- It protects against physical damage and against enemies (it is tough like armour and may be camouflaged)
- As the first point of contact it is sensitive to touch, pressure, hot and cold
- It produces secretions and can excrete waste
- It is self-renewing and self-repairing.

Most mammals have an external covering of hair, with little fat in the deeper layers (cat). Others have little hair and depend on the fat in the deeper layers to keep out the cold (man, pig and whale). Some have sweat glands. Some can make vitamin D in the skin in the presence of sunlight.

The waterproofing and self-renewing properties of the skin depend on the nature of the outer layers (epidermis). No living cells can stand exposure to drying. The outermost cells are dead, and since they are horny and tough they provide protection and cover for the living cells beneath. They are

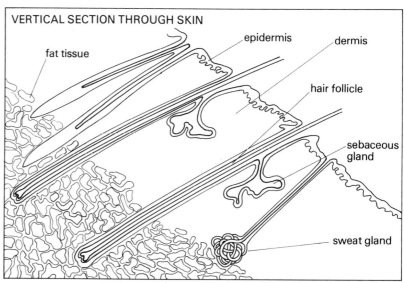
VERTICAL SECTION THROUGH SKIN

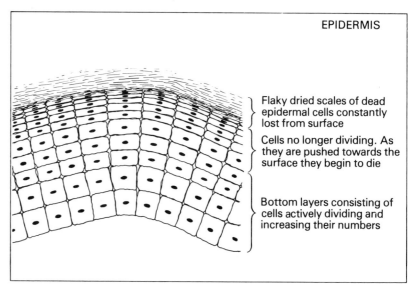
EPIDERMIS

Flaky dried scales of dead epidermal cells constantly lost from surface

Cells no longer dividing. As they are pushed towards the surface they begin to die

Bottom layers consisting of cells actively dividing and increasing their numbers

Principles of Biology

constantly worn away and replaced from below. The thickness of the dead layer varies according to the amount of friction to which it is subjected.

The skin is able to prevent heat loss by the presence of hair. Hair is kept supple by the secretions of the oil glands which discharge into the hair follicle. The hair traps a layer of air which is a good insulator. Fat below the surface is a poor conductor of heat and prevents heat loss.

When the body temperature starts to rise during vigorous activity the skin can get rid of heat. The hair muscles relax and the hairs lie flat, thus decreasing the thickness of the air layer trapped at the surface. Sweat is produced from the sweat glands. As it evaporates from the skin surface it takes up body heat. The flow of blood through the outermost capillaries is increased, causing a greater heat loss by radiation.

Some parts of the skin have a thick horny layer, even at birth. The skin reacts to friction by increasing the thickness of this layer still further.

Further protection is afforded by the development of nails, hooves and claws from the horny layer.

The skin of man is sensitive to touch or pressure. It is most sensitive on the lips and the fingertips. It is least sensitive in the small of the back. It contains sensory cells which react to heat and cold. There are also nerve-endings which signal pain.

All mammals have glands in the skin apart from sweat and oil glands. Mammary glands produce milk in the female. The oil glands in the outer ear tube produce a bitter waxy substance which discourages small animals. Scent glands are usually present. Their secretions may attract members of the opposite sex or they may be

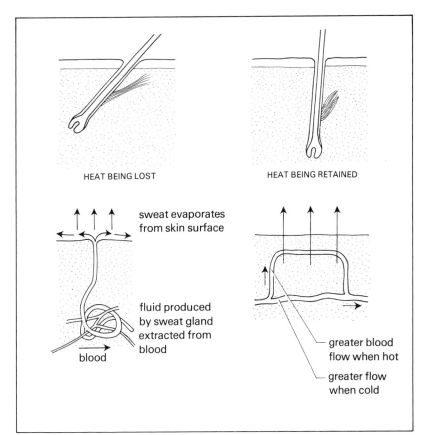

used to drive away enemies, as in the skunk.

In animals which sweat, the skin acts as an organ of excretion, since sweat consists of water containing dissolved urea and salts.

The skin

Support and movement

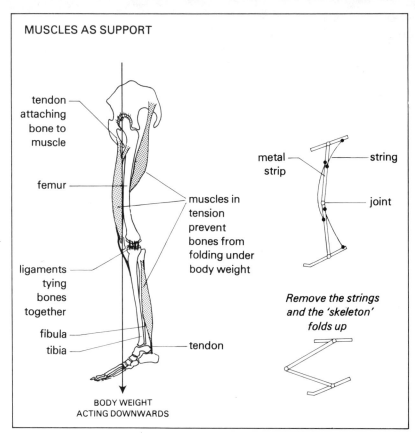

Living cells are soft and flexible. An animal, unless it is very small indeed, cannot support itself and move about on land unless it has hard parts.

Snails have shells. The shell is completely outside the animal. It can increase in size by adding new material at its edge; as the snail grows, the shell grows and protects the snail but plays no part in movement. It is cumbersome and the snail is slow-moving.

Insects have a hard external 'skin' (exoskeleton) which protects, supports and helps in movement. It does not grow, so it has to be shed and replaced by a bigger one at intervals as the insect grows.

All vertebrate animals have an internal bony framework (endoskeleton). Bone has living as well as non-living parts, so it can grow, and even change shape, to keep pace with the growth of the whole animal.

If it were just a matter of support a simple rigid framework would do, like the internal steel framework of a modern building, bolted or welded together. Animals have to be able to move, so their rigid parts must be able to move. Bones are joined together at movable joints – the skeleton is *articulated*. The skeleton alone is still insufficient to support an animal. Other parts are needed to hold bones in position. The support system is a two-part system. It has bones which resist compression (crushing force) and the muscles and tendons which resist tension (stretching force). Support can be compared to the

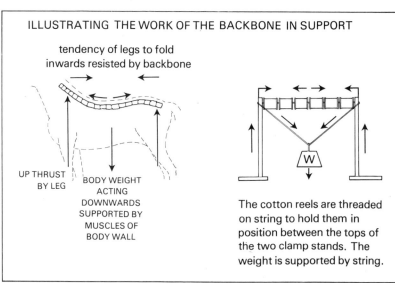

way a tent is kept in shape. The poles provide the rigidity and the adjustable guy ropes keep the poles in position by resisting pulls. At all times, unless you are lying down, completely relaxed, the muscles are doing work, maintaining tension and holding bones in position as the example shows.

A simple model could be made from metal strips, with nuts and bolts representing the joints and string representing the muscles and tendons.

Most mammals walk on all fours. Their weight is supported

106 **Principles of Biology**

on the four columns of the legs with the centre of gravity somewhere between. The backbone has an important part to play. Because it has to be flexible for movement, it is made up of separate parts held together by ligaments. The tendons of the back muscles both above and below the backbone, which cause bending movements in the backbone, also help to hold the whole structure together.

The body weight is in no sense supported on a rigid pole. The only property displayed by the backbone (and the cotton reels in the model) is the ability to resist compression forces along its length. It is able to do this perfectly efficiently although it is completely flexible.

Bone
A living bone is hard and slightly flexible. It is covered with a layer of delicate tissue containing blood vessels, From this layer, tiny blood vessels penetrate the bone to supply the living cells inside it. About half the weight of a bone consists of non-living mineral salts (calcium and magnesium). The rest is made up of living cells and tough fibres which act rather like the metal rods in reinforced concrete. The mineral part of the bone gives it hardness, the fibres give it flexibility.

The mineral part can be dissolved by placing a bone in dilute hydrochloric acid. The bone then becomes soft, rubbery and flexible. Heating a bone strongly in a flame will drive off all the non-mineral part. The result is a completely white bone which is so brittle that it shatters at a touch.

Cartilage (gristle) Some parts of the skeleton require rather more flexibility than bone can provide. In these cases bone is replaced by cartilage. It is bluish in colour and transparent. It is found at the places where the ribs are joined to the breast-bone, in the supporting rings of the windpipe, and in the nose and the ears.

Ligaments Bones are held together at the joints by ligaments. These are made of flexible fibres, very slightly elastic to provide 'give', and extremely tough. They wrap round the joint, rather like a bandage, holding the bones together without seriously hindering their movement.

Tendons are similar to ligaments. They are the 'ropes' which attach muscles to bones. They are flexible but extremely resistant to stretch.

Muscle A single muscle is made up of a large number of fibres all running parallel to each other. The fibres do work by shortening. They need energy to do this, and their requirements of food and energy are high. Much of the body heat results from muscle contraction. The contraction of the muscles attached to the skeleton depends on impulses from the nerve fibres which supply them. Muscles normally act together in pairs, pulling in opposite directions on bones. In this way they keep bones in position. They also move bones, one of the pair creating a much greater pull while the other provides a slight resistance.

A muscle shortens only to a limited extent. The fully contracted muscle is about four-fifths of the length of a relaxed muscle. In order to get maximum movement at a joint, the tendon of the muscle is usually attached close to the joint.

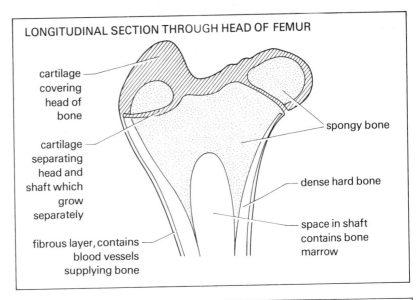

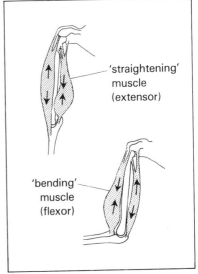

Support and movement

Joints

Some bones are fixed immovably together, for example, the bony plates forming the skull. At movable joints, friction is kept to a minimum – the joint is lubricated and permanently sealed.

Ball and socket (found at the hip and shoulder)
One bone can move in all directions

Many small joints, such as those between the wrist bones, are built like ball and socket joints (one

and-socket type of joint between the skull and the first bone in the backbone (atlas). This allows rocking movements of the head (from side to side and backwards and forwards). The skull, together with the atlas, is able to rotate to some extent (shaking the head) about the

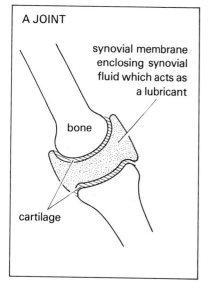

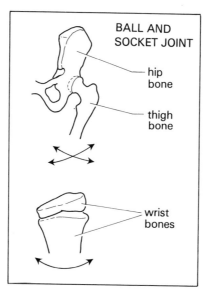

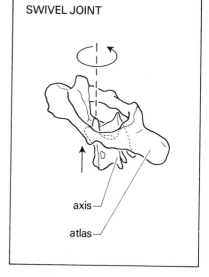

The commonest types of movable joint in the body are:

Hinge (found at knee, elbow, finger and toe joints)
Movement occurs in one plane only

surface convex, one concave) but have more limited movement. The combined action of all of them produces the same sort of total movement as if it were a ball and socket.

The skull moves by means of two joints. There is a flattened ball-

second bone in the backbone (axis). This is because there is a swivel joint between the atlas and axis bones.

Bones and muscles at work
Bones in the body, apart from their ability to provide support, act as simple lever systems with the muscles producing the force to move them. The opposite page shows three simple examples:

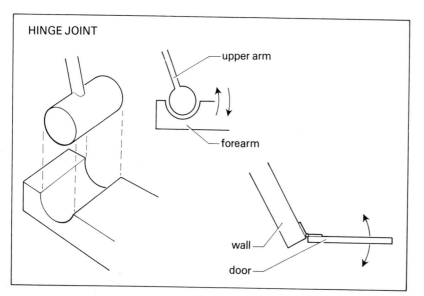

Principles of Biology

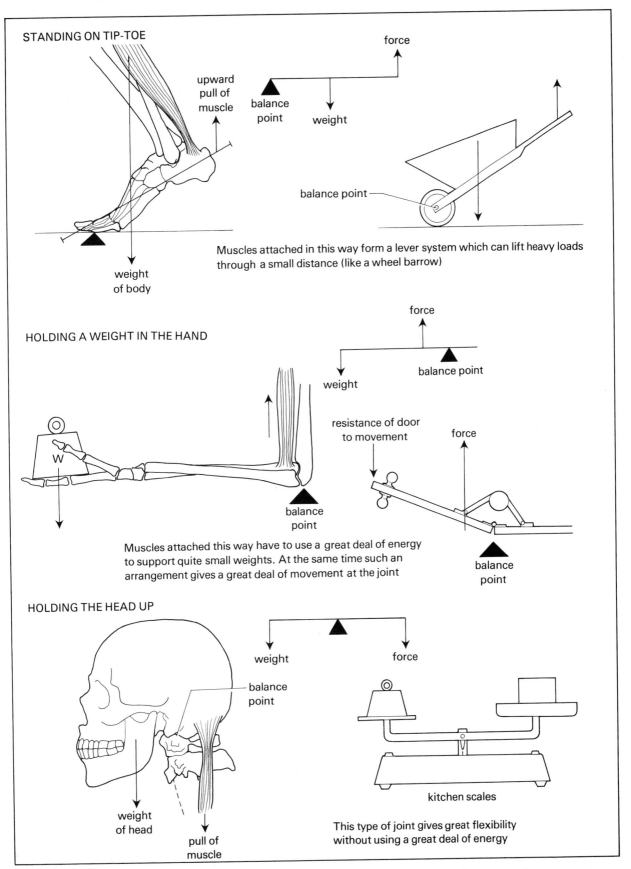

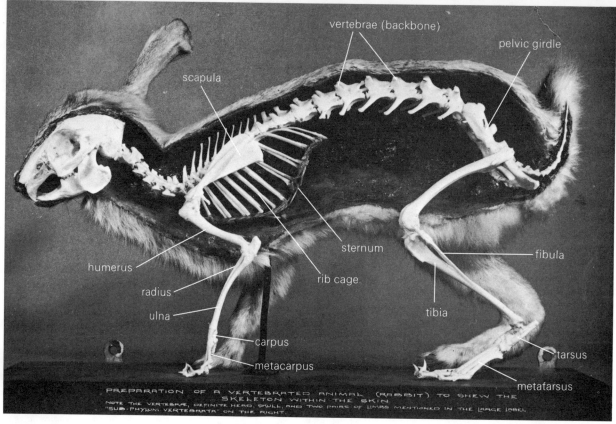

Rabbit skeleton

The skeleton

All mammal skeletons are built on the same common plan:

A backbone, to which the skull and the rib cage are attached. Two sets of limb skeletons together with the limb girdles which anchor them in the body. The shoulder girdle is held in position against the ribs by muscles, tendons and ligaments. The hip girdle is fixed firmly to the backbone because the hind limbs provide the main thrust in movement.

The arrangement in the human skeleton is basically the same but man has two separate bones in the lower part of each arm and leg. In most mammals these bones tend to be joined for greater strength. The arrangement in man allows rotation. We can turn our palms upwards and our feet outwards.

Walking upright on two legs has some disadvantages. The bones in the lower part of the backbone carry a heavier load than the others. This can cause 'slipped disc', when the fibrous pads between the bones are squeezed out and press on the spinal nerves.

Running and turning sharply may cause damage to the knee joint. It carries more weight than it would in a four-legged animal and when a footballer, for example, pivots on one leg the ligaments holding the bones at the knee may easily be torn. This joint is much stronger in other animals because no rotation at the knee is possible and part of the body weight is supported by the front legs.

110 **Principles of Biology**

Human skeleton

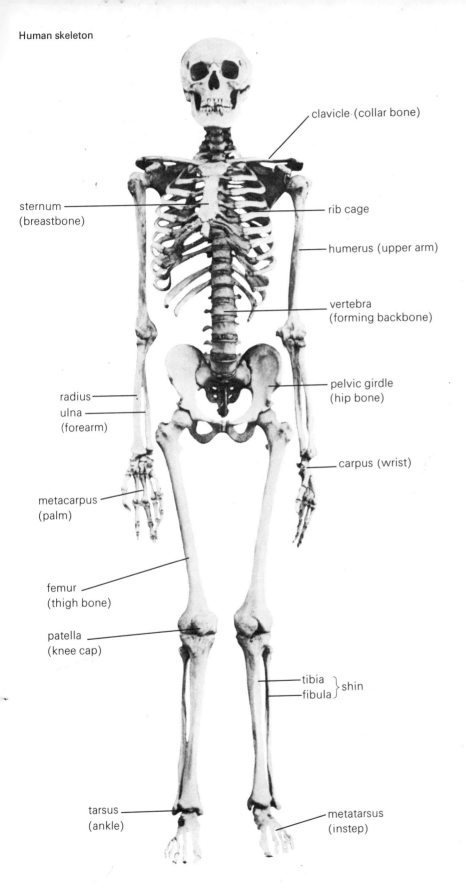

Support and movement 111

Plant and animal behaviour

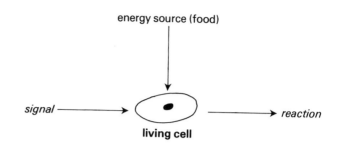

To a biologist, behaviour means all the observable activity of plants and animals. In animals this usually means movement, and movement generally means the action of muscles. Some examples of animal behaviour are: a bird building a nest, a boy riding a bicycle, or a kitten playing with a ball of wool. Plants also move and, if you have patience enough, it is possible to watch a flower closing its petals at nightfall, or a young shoot turning towards the light.

Behaviour sometimes seems simple (*Amoeba* moving away from a bright light) and sometimes complex (a man writing a letter). *Amoeba* is a single-celled animal and its behaviour is the activity of a single cell. Man is a multi-celled animal and his behaviour is due to the activities of a large number of cells. So, in order to understand the behaviour of complex organisms, we must begin by trying to understand something about the activities of single cells.

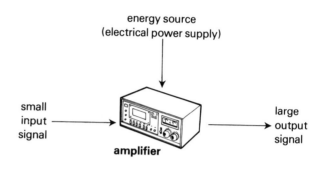

Irritability
All living cells react to changes in their surroundings. This ability is sometimes referred to as *irritability*. In order to react to a change, or signal, the cell requires energy. This it gets by the respiration of food materials. An electronics engineer would say that a cell is rather like a miniature amplifier. An amplifier transforms a small signal into a larger one and it must have a power supply in order to do this. Cells sometimes amplify signals

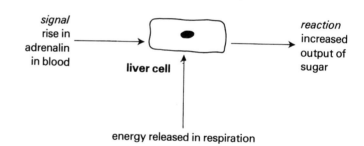

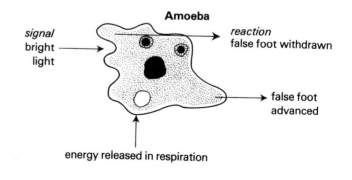

Principles of Biology

but they are also able to alter them to produce a different kind of signal.

Animal behaviour

Although all animal cells have the ability to react to changes in their surroundings, the activities of the whole organism depend on the properties of three special kinds of cell. They are sensory cells (receptors), nerve cells (transmitters) and muscle or gland cells (effectors).

Sensory cells are found in the skin, in the special sense organs such as eye and ear, and in the muscles and joints. There are various kinds reacting to light, heat, pressure and so on, but in each case the reaction causes activity in a nerve cell.
Sensory cells transform signals.

Nerve cells are sensitive to chemical signals. They react by generating a different kind of signal which is called a nerve impulse. **Nerve cells transmit and distribute signals.**

Muscle and gland cells react to nerve impulses. Their reaction is mechanical. Muscle cells shorten, gland cells produce secretions.
Effector cells transform signals.

All three types of cell work together in the reactions of the animal to the world around it. Sensory cells pick up signals from outside and change them. The changed signals are then carried by nerve cells to the effector cells. Here they are transformed into action in the form of muscle contraction or gland secretion. Each cell in the chain shows irritability by reacting to signals. When we observe the whole animal, we see only the original signal and the final reaction, such as bright light causing the pupil of the eye to get smaller. The whole organism reacts to its surroundings. The signal is called a *stimulus*; the reaction is called a *response*.

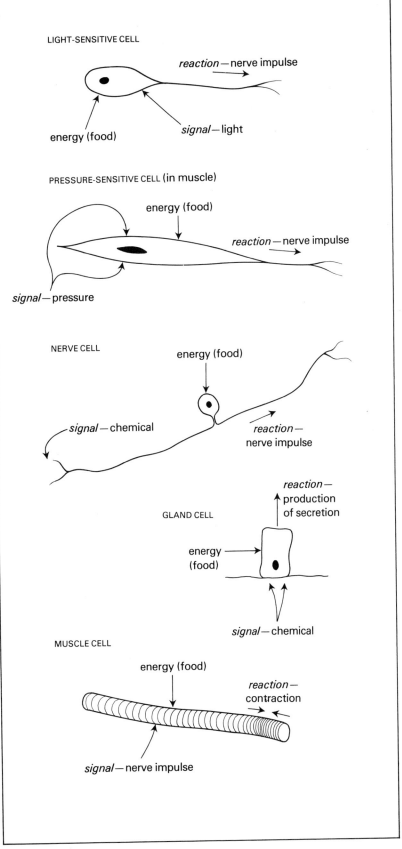

Plant and animal behaviour

The nervous system

The nervous system in a mammal consists of the brain, the spinal cord and a system of branching nerves extending to every part of the body. It consists of vast numbers of simple units called nerve cells. These conduct signals called nerve impulses.

The main task of the nervous system is to provide a *communication* system between the sensory cells and the muscles and glands. In this way it is like a telephone network with the brain and spinal cord acting as the central exchange. It also *co-ordinates* the actions of muscles and glands. The 'wiring' of the system ensures that the muscles as a whole work together. Computers, too, are made up of large numbers of simple conducting units joined together. A computer is a machine for storing information and solving problems. The nervous system is far more complicated than any man-made computer. It contains thousands of millions of nerve cells. It also *stores information* and *solves problems*.

Nerve cells

A nerve cell reacts to a chemical signal supplied at one end, by generating a nerve impulse. It travels along the nerve fibre at speeds up to 100 metres per second (m/s). A nerve impulse is a kind of electrical disturbance which leaks its way along the fibre. When it reaches the end of the nerve fibre it produces a chemical substance. The chemical reaction produced at the end of one nerve fibre can act as the signal for another, giving chains of connected nerve cells. The links between nerve cells are called synapses. A signal can travel one way only across the synapse. The synapse is also a resistance. It may need a number of impulses arriving at a synapse before the signal can 'jump the gap'. A synapse is a junction; it is a point where at least two nerve cells meet.

Nerve cells all work in the same way. All nerve impulses are alike. The cells may have different shapes, depending on the sorts of connections they make.

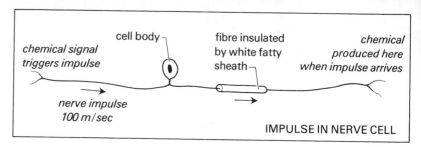

IMPULSE IN NERVE CELL

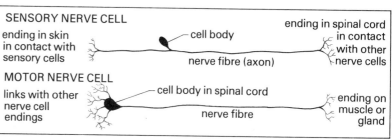

Reflex action

Complex patterns of activity are made up of simpler bits, just as a large building is made up of simpler pieces. The simplest bits of behaviour are reflex actions. A reflex action is a fixed response to a stimulus. The same stimulus

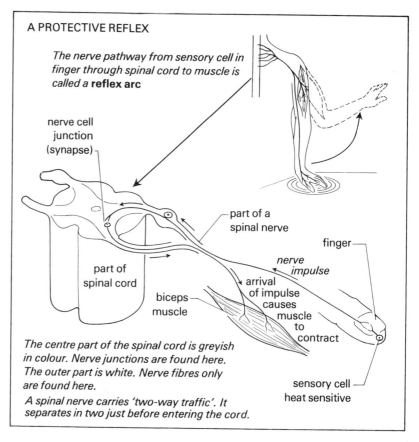

A PROTECTIVE REFLEX

The nerve pathway from sensory cell in finger through spinal cord to muscle is called a **reflex arc**

The centre part of the spinal cord is greyish in colour. Nerve junctions are found here. The outer part is white. Nerve fibres only are found here.

A spinal nerve carries 'two-way traffic'. It separates in two just before entering the cord.

Principles of Biology

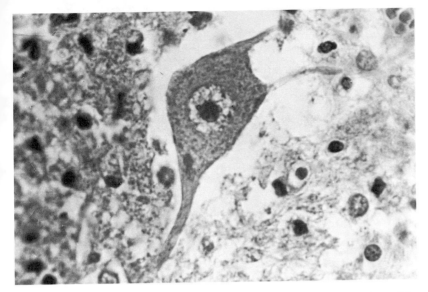

Photomicrograph of a motor nerve cell

A nerve fibre and motor ending on a muscle fibre

always causes the same response. Bright light shining in the eyes causes the pupils to get smaller. In dim light they get larger. When the skin is cold, goose pimples appear. When food is eaten, saliva flows. Dirt in the eye causes it to water. These are examples of reflex action. Reflexes are automatic; they do not have to be learned. They may be protective or they may control. When the doctor tests your knee reflex, he is testing those reflexes which come into action when a person is standing. The muscles do their work without the person having to think about it. Any tendency of the leg to fold under the body weight automatically causes the muscle to counteract this. All reflexes have the same basic structure.

Why do the nerve fibres from the sensory cells in the skin have to travel all the way to the spinal cord? Why do they not connect directly to the muscle? There are many good reasons. Your ear may be itching and you will need to bend your arm to scratch it. There are hundreds of occasions when the biceps muscle must contract to bend the arm and the signals could be coming from almost anywhere. The spinal cord acts as the central exchange for all signals destined for the biceps muscle. It also amplifies the signal since a small number of nerve fibres from the tip of the finger could not possibly bring all the separate fibres in a large muscle into action at the same time.

The more complicated the behaviour is, the more complicated the 'wiring' is. In most animals the brain has complex circuits which control instinctive behaviour, such as nest building or courtship. Such activities are made up of many reflex actions which are interlocked, and dependent on signals from outside the animal. Instinctive behaviour is not learned. It is inherited, like eye colour.

The most intelligent animals, especially man, are capable of much more flexible behaviour.

Man has no clearly instinctive activity at all. He acquires a programme of behaviour as he goes through life. He has to learn by experience and he can do this

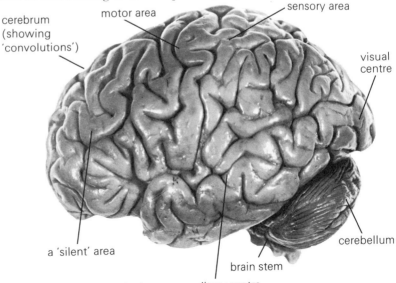

Side view of entire human brain

Plant and animal behaviour 115

A modern computer (IBM system/370 model 168)

because he can remember. A simple animal is in some ways like a pocket computer. The animal has a number of fixed reactions to a limited number of situations. The computer can do a limited number of operations quickly and efficiently. On the other hand, the brain of man not only reacts to the signals feeding into it, but is actually altered and affected by the signals. Some change in its circuitry takes place. To a very limited extent this is possible with some of the most complicated and highly expensive modern computers.

Chemical co-ordination

The nervous system forms a very effective system for transmitting messages from sense organs to muscles, but it is not the only communication system in the body. Many processes in the body are controlled by messages in the form of chemical substances (hormones) circulated in the bloodstream and produced by special glands, the ductless glands. Hormones produce effects which tend to be long-lasting or continuous. It is obviously much simpler to control, for example, the *rate* at which body cells work by maintaining a steady level of a particular chemical substance in their blood supply than for every single cell to be supplied with its own private 'telephone' line. Furthermore, nerve fibres can produce only one chemical at the nerve ending, so that only one message is possible. Different chemical substances circulated in the blood may produce different effects in the same structure or organ.

Ductless glands do not act independently. They often interact, or they may produce opposing effects, or one gland may exercise control over another through the blood. Most of the ductless glands are directly controlled by the pituitary, which is often called the master gland. It is in close contact with the brain.

The pituitary interacts with other glands as shown by the arrows.

Apart from producing hormones which affect the other glands the pituitary produces a 'growth' hormone. An overactive pituitary

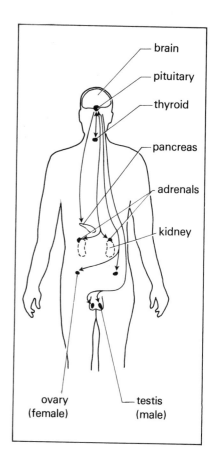

116 **Principles of Biology**

can cause excessive growth; if it is underactive during childhood, the person remains small.

The adrenal gland produces several hormones including adrenalin. Adrenalin is produced especially when the person is frightened or angry. It causes a rise in heart rate, blood pressure, breathing rate, blood sugar level.

It brings the body to the state of 'fight or flight'.

The pancreas produces insulin. Insulin causes sugar to be removed from the blood and stored in the liver. Lack of insulin in the blood causes a high blood-sugar level and its loss through the kidney (diabetes). Insulin's effect on sugar level is the opposite of that of adrenalin.

Sense organs

We have seen how the brain as a whole has capabilities which are not possessed by its separate nerve cells. Its powers result from the way in which its vast numbers of cells are organised. This is also true for the special sense organs of the body.

An earthworm has large numbers of light-sensitive cells scattered throughout its skin. The earthworm is therefore sensitive to light. We also have light-sensitive cells but ours are collected together in a special organ, the eye. The eye organizes the light entering it so that it gives information about the outside world. The light reflected from objects tells us their position, size, shape, distance and colour. This is what we mean by *seeing*. The worm, in spite of its light-sensitive cells, cannot see.

The eye

In many respects the eye is like a television camera. It can be pointed in various directions by the six external muscles attached to the wall of its protecting bony socket. It has a light-sensitive surface (retina). It has devices for forming an image. Light can enter only through a small hole at the front. The size of the hole (aperture) can be altered. The eye, like the TV camera, is relaying information continuously. The camera sends its information by cable to the transmitting station as a series of electrical impulses; the eye sends its information to the brain as a series of nerve impulses.

The retina is a surface which converts light energy into nerve impulses. The pattern of nerve impulses in the optic nerve varies according to the pattern of light falling on the retina. This pattern is called an image.

A clear glass marble, a raindrop or a round flask filled with water can form an image. If you hold a plain sheet of paper up to a window in daylight and place the water-filled flask in front of it, an image of the window will appear upside-down on the paper. The whole eyeball also behaves in the same way, as if it were a glass marble, although of course the eyeball has a lightproof cover over most of its surface. If you are able to get hold of a fresh ox eye, cut a 'window' in the rear surface, cover this with a piece of greaseproof paper and then hold it up to the light. You will see on the paper an upside-down image of the window on the paper. The sharpest image is formed just inside the eyeball, in fact where the retina is, and not behind the eyeball, as is the case with the flask. This is because the surface of the eyeball is more sharply curved at the front where the light enters and it has another lens inside it. The curved front surface, the watery fluid and the jelly-like material all help to form the image. The extra lens inside is used mainly for adjusting the focus for near objects.

Light from a distant point strikes the eye as a parallel beam. The eye is a *converging* lens. It brings the light back to a point on the retina.

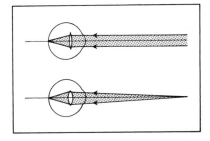

Light from a near point strikes the eye as a spreading beam. The eye must converge the light still more to bring it to a point on the retina. The internal lens alters its curvature, becoming fatter. This change is called *accommodation*.

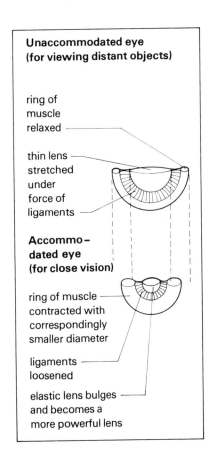

Unaccommodated eye (for viewing distant objects)

ring of muscle relaxed

thin lens stretched under force of ligaments

Accommodated eye (for close vision)

ring of muscle contracted with correspondingly smaller diameter

ligaments loosened

elastic lens bulges and becomes a more powerful lens

Plant and animal behaviour

The human eye has two common faults.

Short sight: a short-sighted person has a lens system which is too powerful. A magnifying glass is useful only at close range and this is also true of the eyes of a short-sighted person. A diverging lens corrects for this.

Long sight: a long-sighted person has eyes which have the opposite fault. Objects at a distance can be seen clearly but not those at close range. These remain blurred because the lens cannot converge

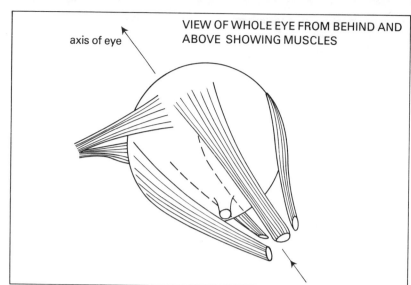

VIEW OF WHOLE EYE FROM BEHIND AND ABOVE SHOWING MUSCLES
axis of eye

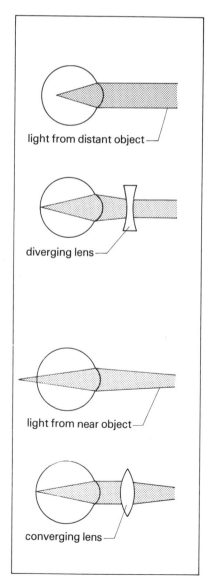

light from distant object

diverging lens

light from near object

converging lens

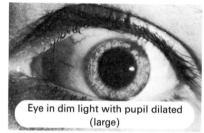

Eye in dim light with pupil dilated (large)

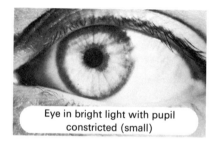

Eye in bright light with pupil constricted (small)

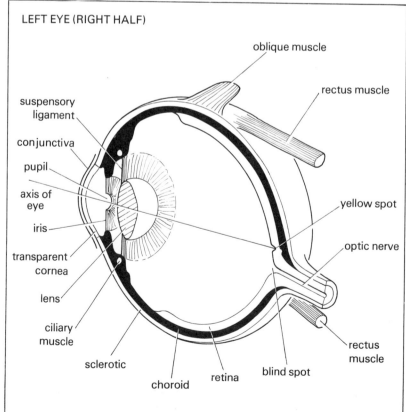

LEFT EYE (RIGHT HALF)

oblique muscle
rectus muscle
suspensory ligament
conjunctiva
pupil
axis of eye
iris
transparent cornea
lens
ciliary muscle
sclerotic
choroid
retina
blind spot
yellow spot
optic nerve
rectus muscle

118 **Principles of Biology**

the light rays enough. A converging lens corrects for this.

A different kind of long sight comes with increasing age. The lens in the eye stiffens and fails to accommodate. A converging lens must be worn for close work.

The ear

The ear is perhaps a difficult organ to understand. One reason for this is that the works are buried deep in solid bone and they are not symmetric like the eye's. Cut an eye in half and everything can be seen. Try to imagine a bicycle dropped into a bath of concrete which then sets hard. If you did not know what a bicycle looked like it would be very difficult to work out its structure just by cutting the block of concrete in half. Another reason for difficulty is that the ear is not just one sense but three. Confusion is added because, in ordinary language, we refer to the flap on the outside as the ear. To a biologist, the flap is a very small part of the ear.

The sensory cells in the ear are all alike. Each has tiny hair-like projections. Anything which bends or pulls on the hairs causes a reaction in the cell and a signal to be

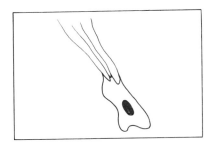

generated in a connected nerve fibre leading to the brain. The complex structure of the ear is simply an arrangement which decides what external changes will cause such bending or pulling to take place. In order to explain how the ear works we shall not worry too much about the detailed structure. The way in which it is fitted into the skull wall is largely a matter of space-saving. It is much more important to understand its working.

Sensitivity to movement of the head

Each ear contains three curved, fluid-filled tubes (semicircular canals). They are arranged at right angles to each other. The patches of sensory cells have their hairs embedded in a tiny lump of stiff, jelly-like material floating in the fluid. A sudden movement of the head causes the jelly to get momentarily left behind. Since the cells are fixed, the hairs get bent and a signal is generated in the attached nerve fibre. The pattern of nerve impulses set up in the three different canals depends on the direction and force of the movement.

Sensitivity to direction of gravity

Part of the ear consists of a chamber (utricle) lined with sensory cells. A small particle of calcium carbonate is contained in this chamber and is free to fall in any direction under the influence of gravity. Wherever it falls, it bends the hair cells, and nerve signals are set up. Different signals are generated according to whether the person is upright, standing on his head or lying down. The brain can then act, if necessary, to restore balance.

All vertebrates possess these structures for detecting movement and direction of gravity. They are vitally important in fish: imagine the problems of a herring in the sea on a dark night. They are basically navigational aids: very complex devices, based on the same principles, are used in modern submarines. They are not so important for land animals like mammals, though peculiar effects are produced when they fail.

Hearing

Land animals can hear sounds and interpret them. Sound is caused by waves of pressure spreading through the air like ripples on a pond. The greater the frequency of the waves, the higher the pitch. The human ear is sensitive to only a limited range of variation in frequency just as the eye is sensitive only to a limited frequency range of radiation, called the visible spectrum. The hearing part of the ear (cochlea) sorts out the different frequencies of sound waves reach-

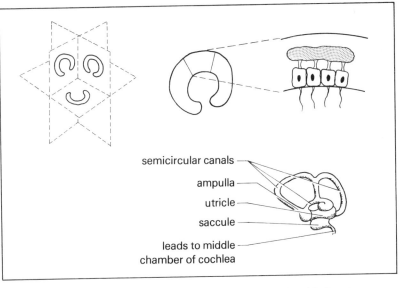

Plant and animal behaviour

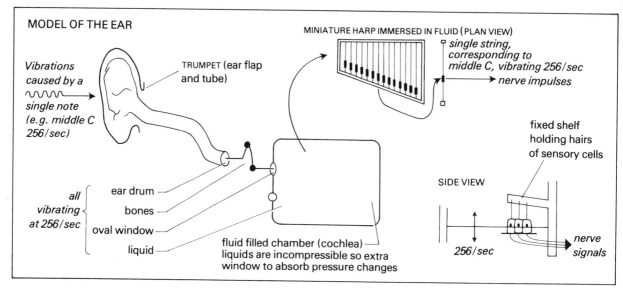

ing the sensory cells. Just as the eye sorts out images and converts them into patterns of nerve impulses in the optic nerve, so the cochlea in the ear sorts out different sound frequencies and converts them into patterns of nerve impulses in the auditory nerve. The diagram above represents the basic working arrangement of the hearing part of the ear.

The ear flap serves as a direction finder in those animals which can move it. In man it serves no useful purpose. The tunnel leading to the eardrum protects the drum, which could work just as well on the surface, as it does in a bird.

The eardrum vibrates in response to vibrations of the air, in the same way as the thin metal disc (diaphragm) in a microphone.

The vibrations are transferred across the air-filled middle-ear by the three tiny ear-bones to the oval window in the wall of the skull. The reason for such an arrangement is fairly simple. The oval window is very small, and can vibrate only through a much smaller distance than the eardrum. It has a fluid in contact with it on its inner surface. Vibrating air simply has not enough energy to vibrate it. We know that this is so because people with damaged eardrums or ear bones are deaf. The bones act as a simple lever system.

The vibrations are finally transferred to a liquid-filled cochlea, by the oval window.

The cochlea The pattern of air vibrations we call sound are reproduced in the liquid in the cochlea. The cochlea contains a different type of diaphragm or membrane. It is like a miniature harp. When one part (single string) of a real harp is made to vibrate it can pro-

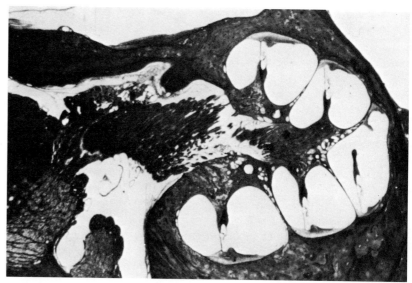

The cochlea in section

duce only one note. Each part of the cochlea 'harp' can vibrate only in response to a particular frequency (note). The sensory cells that are vibrated and the pattern of nerve impulses created depends on which part of it vibrates. The cochlea is a miniature 'wave analyser'.

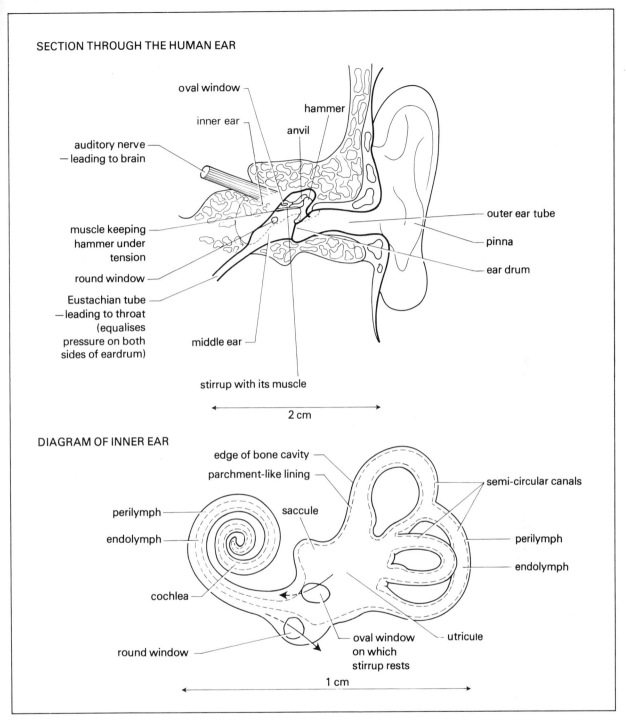

Plant behaviour

Plants respond slowly to stimuli (much slower than animals), they respond to fewer stimuli and, in flowering plants, the responses are made only by parts of the plant. There are some exceptions, but plant responses are generally growth responses. This means that responses can be made only in the growing parts of the root, shoot and leaf. The changes to which flowering plants react are like instructions about the direction in which to grow. Such reactions are called *tropisms*. The plant organ's growth is controlled by the direction of the stimulus. Not all plant reactions are tropisms. The petals of some flowers close up at night. If a mimosa plant is tapped, its leaves droop. These are *nastic* reactions.

Plant stems are mostly positively

Plant and animal behaviour

phototropic, which means that they grow towards light. They are negatively geotropic which means that they grow away from the direction of gravity. Roots are very little affected by light. They are positively geotropic which means that they grow towards the earth.

Experiment to find the effect of light on stems

Place some wet cotton wool in three dishes and sprinkle cress seeds in each. When the germinated seedlings are about 2 cm high, place one group in complete darkness, expose the second group to continuous all-round light and place the third group in a box with an opening only at one side. After a few days the results are as shown in the diagram below.

The stems in the dark grow faster than the stems in the light. The ones in group 3 have grown faster on the dark side than on the light side, and have therefore turned to the light.

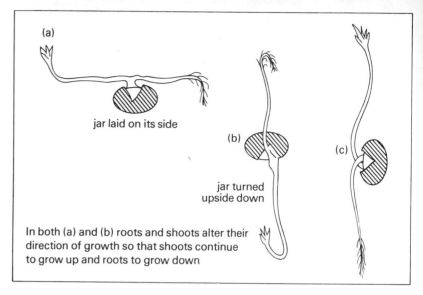

In both (a) and (b) roots and shoots alter their direction of growth so that shoots continue to grow up and roots to grow down

Experiment to find the effect of gravity on stems and roots

Take three jam jars and line them with blotting paper. Fill the jars with damp sawdust. Push two or three soaked bean seeds in each jar between the glass and the paper. Place the jars in a dark airy cupboard for a few days. When the seeds have germinated and both shoots and roots are well-established turn one jar (a) on its side and another (b), upside-down. The results are as shown above.

A growth-promoting substance in the stem is produced just below the tip; if the tip is removed, growth slows and the stem fails to react to light. This substance, auxin, is soluble and diffuses down the stem. It is partly destroyed by light. This is why growth is slower in the light, or on the lighted side in stems exposed to one-sided illumination. When the stem is laid on its side, the auxin collects on the lower side which is also the darker side. Growth is therefore speeded up on the lower side.

Auxin is also produced in the root tip, but here it *slows* growth. (Roots and stems have different sensitivities to auxin. In living organisms, reactions often reverse at different concentrations of materials. Some drugs are beneficial in low doses but become poisons at higher doses, for example, aspirin.) When the root is laid on its side the auxin collects on the lower side where it slows growth. The root therefore grows downwards.

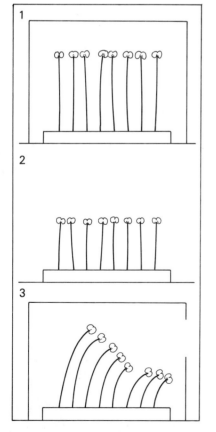

Reproduction

Animals and plants do not live for ever. Many are eaten by organisms higher in the food chain. Others are destroyed by cold, drought or lack of food. Those which escape have a self-destruction system; they die of old age. They are replaced by a process called *reproduction*.

Allowance is made for the fact that life can be even more difficult for the young; for example, a female cod lays several million eggs each year, of which only a relatively small number hatch successfully.

Most organisms reproduce by *sexual* reproduction. They produce reproductive cells (gametes) of two kinds – female (egg) cells and male cells. Egg cells are generally larger than most ordinary body cells and contain stored food. The male sex cells are very much smaller. In animals, they are able to move independently in search of the egg by means of a whip-like tail. In plants they are not able to move and must depend upon something else to carry them. Most female gametes do not need to move.

A new individual begins life when a male and a female cell fuse together. The two sorts of gamete may be produced in the same organism, as is the case for most plants, or in different male and female organisms, as is the case in all mammals. Sexual reproduction is a way of maintaining variety in a species, since the offspring are usually different from their parents.

Reproduction in flowering plants

Flowers are the organs of sexual

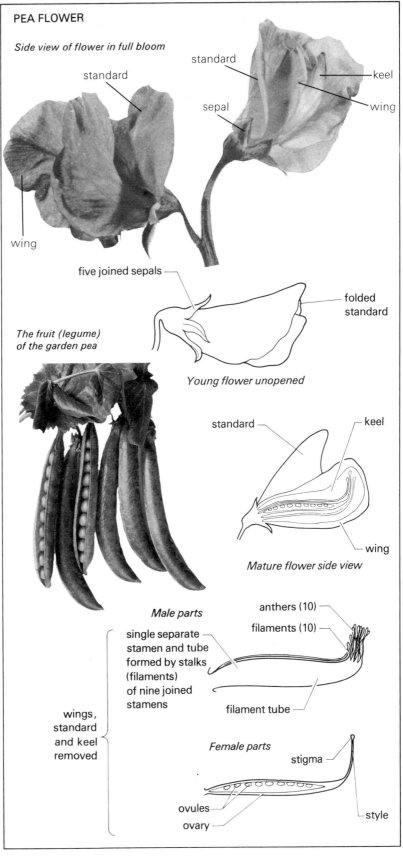

PEA FLOWER

Side view of flower in full bloom

The fruit (legume) of the garden pea

Young flower unopened

Mature flower side view

Male parts — single separate stamen and tube formed by stalks (filaments) of nine joined stamens

wings, standard and keel removed

Female parts

Reproduction 123

reproduction in flowering plants. A flower is a special kind of shoot, complete with stem and leaves. It starts as a bud, often indistinguishable from other buds. The pea, like most flowers, has four kinds of 'leaf'. Starting from the outside they are the sepals, petals, stamens and carpels. It has:

Five green *sepals*, all joined, protect the flower in bud.

Five *petals*, usually white, attractive to insects (usually bees) which visit flower in search of nectar produced by *nectaries* near base of petals. Two petals are joined to form a boat-shaped 'keel' containing stamens and carpels.

Ten *stamens*, each consisting of a hollow, lobed case (anther) supported by a slender stalk (filament) – anther contains pollen grains, very resistant spores, protected by an outer woody case.

Single *carpel*, enclosed by tube, formed from flattened, joined, lower parts of filaments. Carpel resembles a green leaf folded about its midrib; this extends beyond the leaf, forming a stalk (style), ending in the stigma which acts as a landing platform for pollen. Carpel contains a number of ovules each with an egg cell.

Pollination
Pollination is the transfer of pollen grains from anther to stigma. This must happen before the fusing together of male and female sex cells (fertilization) can take place. Peas are pollinated by insects. The keel is pressed down by the weight of a bee landing on it and this exposes the anthers and stigma. As the bee brushes against the ripe anthers, pollen is rubbed off on to its body, and this, or pollen from another flower previously visited by the bee, gets brushed on to the sticky stigma.

The pollen grain case acts as a protective capsule in which the male sex cell is carried to the female part of the flower. The pollen grains grow, stimulated by chemical secretion from the stigma; a tube emerges through a pore in the pollen grain case, grows down the style and through the wall of the ovary, finally penetrating an ovule. The male sex cell is carried in the tip of the pollen tube and fuses with the egg cell. Fertilization is now completed.

After fertilization, all the flower parts wither away except for the ovary and the ovules. These continue to increase in size. The ovules are transformed into seeds containing a small embryo plant and a food supply. The ovary wall becomes the fruit. It gradually dries, splits, twists and flicks the seeds away from the plant. The pea is an annual plant and dies away in the autumn. Only its seeds survive to germinate in the following spring.

Other plants
The garden pea illustrates one common type of flower. Others belonging to the same family, with a similar structure, are sweet pea, runner bean, gorse, broom and laburnum.

Flowers vary in every way; in the number of parts, in shape, size, colour and scent. Brightly-coloured flowers are pollinated by insects which visit them in search of nectar or pollen. Many flowers are wind-pollinated, particularly trees and grasses. They usually lack sepals and petals which might hinder the dispersal of pollen. They have large stamens and feathery stigmas hanging well clear of the flower. Their pollen is light and powdery and is produced in large quantities. The flowers are almost always gathered into clusters, forming catkins on many trees. Wind-pollinated trees produce their flowers before the leaves appear, while grasses carry their flowers on long stalks well above the rest of the plant.

A single dandelion flower (floret) (A), Dandelion – the 'flowers' are really a cluster of flowers (B)

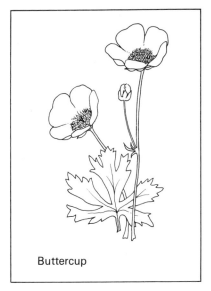
Buttercup

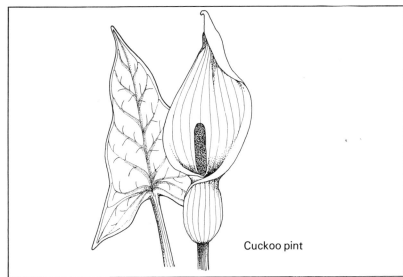
Cuckoo pint

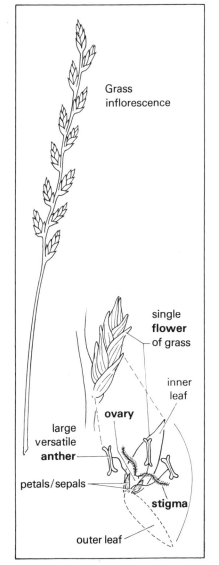
Grass inflorescence

single **flower** of grass
inner leaf
ovary
large versatile **anther**
petals/sepals
stigma
outer leaf

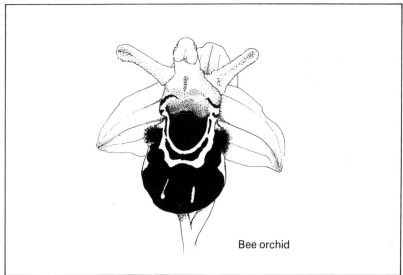
Bee orchid

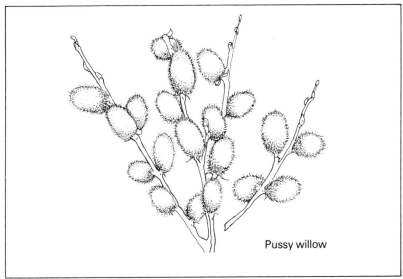
Pussy willow

Reproduction

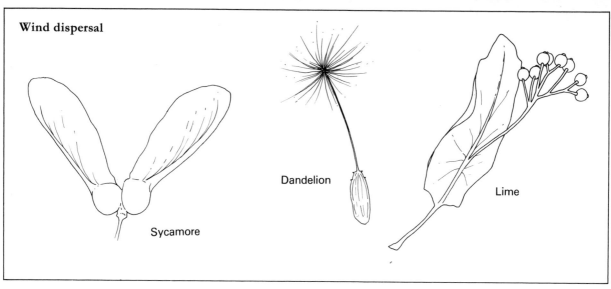

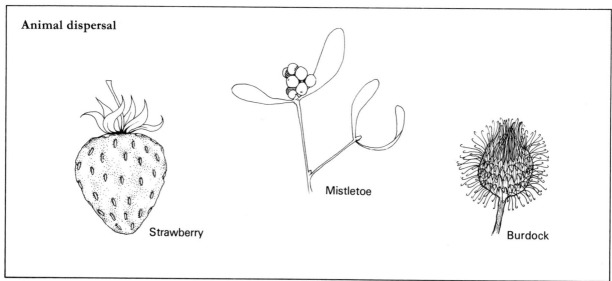

The dispersal and growth of seeds
Seeds must be spread away from the parent plant so that parent and offspring do not compete for living space and the necessities for survival. The fruit, formed from the ovary, is the chief means of seed dispersal although other parts may sometimes be involved. Fruits may be dispersed with the aid of the wind, or be carried on the inside, or outside, of animals. They may also develop some means of throwing out the seeds with enough force to carry them clear of the plant. The seeds of water plants, or of those living close to water, are often spread by water currents.

Most seeds do not begin to grow as soon as they are dispersed. Before they grow they must go through a period of exposure to low temperature. The importance of this is clear. The rate of growth of plants is influenced by temperature and many are killed by cold. If seeds germinated in autumn or late summer they would be destroyed by the cold of winter. After the seed has passed through this 'dormant' period, it will germinate if water and oxygen are present and if the weather is

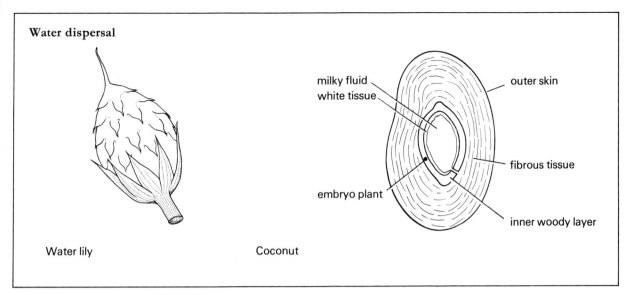

Water dispersal

Water lily

Coconut
- outer skin
- milky fluid
- white tissue
- fibrous tissue
- embryo plant
- inner woody layer

warm enough. An experiment to show this need for warmth, water and oxygen is illustrated here: Test tubes A, B and C are put in a warm place while D is put outside, if the experiment is done in winter, or in a refrigerator. A is the only test tube satisfying all the conditions and germination will occur only in A.

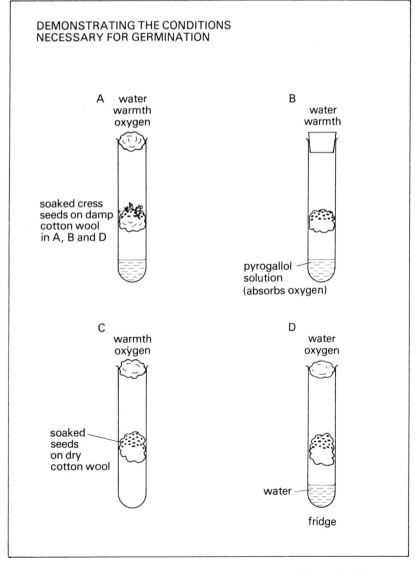

DEMONSTRATING THE CONDITIONS NECESSARY FOR GERMINATION

A water, warmth, oxygen — soaked cress seeds on damp cotton wool in A, B and D

B water, warmth — pyrogallol solution (absorbs oxygen)

C warmth, oxygen — soaked seeds on dry cotton wool

D water, oxygen — water, fridge

Reproduction

The development of seeds

This may be illustrated by the pea. Inside the seed the young plant is attached to two large swollen structures which are the food store, consisting mostly of starch, with some protein. These structures are called seed leaves (cotyledons); they also contain enzymes which will digest the food when growth begins.

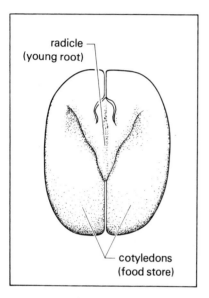

Pea seed with outer covering (testa) removed showing food store (cotyledons) and young root (radicle), and embryo

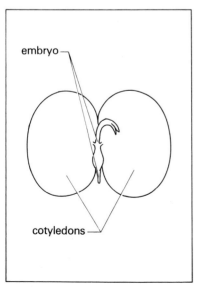

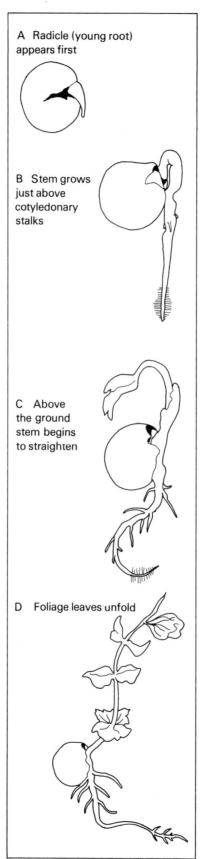

A Radicle (young root) appears first

B Stem grows just above cotyledonary stalks

C Above the ground stem begins to straighten

D Foliage leaves unfold

Vegetative reproduction and food storage

Many plants die away as winter approaches, leaving only protected seeds to survive the winter. Perennials, plants which live for several years, partly die down in autumn. Many trees shed their leaves; non-woody plants lose their above-ground growth. In each case the food necessary to support growth in the following spring is stored somewhere in the plant. Trees store it in their woody stems. Others store it in roots, underground stems or leaves.

Other plants are able to spread rapidly in summer. The strawberry puts out fast-growing horizontal stems which spread in all directions. At intervals along these stems, buds and roots appear which grow into young plants and become detached from the original plant. This is really a form of *asexual* (non-sexual) reproduction.

Man takes advantage of both properties. He may use the stored food directly, as he does when he eats potatoes. He may take advantage of the fact that parts of plants, usually stems with food stores, are able to regenerate a complete plant exactly like the original. He may make use of the way the strawberry plant propagates itself, to increase his stock of plants, for food or pleasure. In all cases the new plants are exactly like the parent plants, so that special varieties, obtained by crossbreeding, can be maintained in a way that is not possible by seeds.

In these cases man is exploiting the natural capacities of plants. The grafting of fruit trees is an artificial application of this property – it could not occur in nature.

128   Principles of Biology

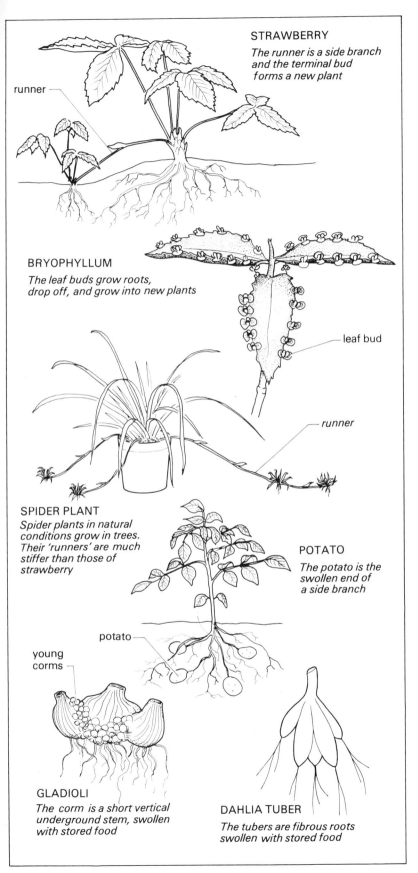

STRAWBERRY
The runner is a side branch and the terminal bud forms a new plant

runner

BRYOPHYLLUM
The leaf buds grow roots, drop off, and grow into new plants

leaf bud

runner

SPIDER PLANT
Spider plants in natural conditions grow in trees. Their 'runners' are much stiffer than those of strawberry

POTATO
The potato is the swollen end of a side branch

potato

young corms

GLADIOLI
The corm is a short vertical underground stem, swollen with stored food

DAHLIA TUBER
The tubers are fibrous roots swollen with stored food

Human reproduction

In its basic details human reproduction is not very different from that of mammals generally.

The ovaries produce the eggs and contain eggs at different stages in development. An egg is shed at approximately 28-day intervals. As it drifts slowly down the egg duct (oviduct or Fallopian tube) it may fuse with a sperm cell, if sexual intercourse (copulation, mating) has occurred. The fertilized egg reaches the uterus (womb), an enlarged part at the end of the oviduct. It embeds itself in the wall of the uterus, which has a very rich lining of blood vessels.

After about three weeks, the foetus, as the new individual is called, is clearly recognizable and has formed all its main organs, although it is only 1–2 cm long. Its skin, at the navel, grows out and folds back over the body so that the foetus is enclosed in a protective, fluid-filled bag. The umbilical cord also arises from the navel, connecting foetus with placenta (afterbirth) which is in close contact with the uterine wall. The placenta contains many blood vessels, connected to the foetus by an artery and a vein passing through the cord. Oxygen and dissolved foods diffuse from the mother's blood into the blood vessels of the placenta. Waste materials, such as carbon dioxide, diffuse back into the mother's blood. The placenta also acts as a protective barrier preventing harmful materials from passing to the foetus.

The foetus increases in size and the uterus stretches to keep pace with it. After about nine months the baby will be about 50 cm long and weigh about 3 kg. The uterus wall begins to contract regularly, slowly at first and then with increasing force and frequency. The baby moves head-first through

Reproduction

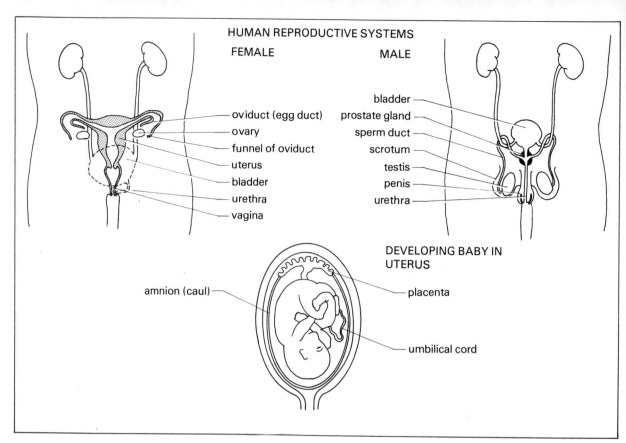

the neck of the uterus and into the birth canal (vagina). It finally passes out of the mother through the opening from the vagina (vulva). The uterus contracts still more, forcing out the placenta. The umbilical cord is cut and it and the placenta are disposed of. The baby is usually fed by milk from its mother until it is several months old, but bottle-feeding from birth is common.

The male has paired glands (testes) corresponding to the ovaries, but they are contained in a pouch (scrotum) outside the body. The sperm ducts lead into the abdomen where they are both joined to a small muscular pouch. From this, a single tube called the urethra leads to the outside. The urethra also carries the outflow from the bladder. The final part of the urethra passes along the centre of the penis, which is made of a spongy tissue with lots of blood spaces.

When sexual intercourse is about to take place, the penis enlarges and becomes rigid because more blood flows into it than leaves it. It is placed in the vagina of the female and semen (fluid containing sperm produced by the testes) is forced by muscular contraction out of the penis. The sperm are not active in the testis. A secretion from a gland called the prostate activates them as they pass down the urethra and they remain active inside the female for several hours. They swim through the uterus into the oviduct. The first sperm to meet the egg enters, shedding its tail as it penetrates. No other sperm cell is able to enter the egg after fertilization has occurred.

Puberty
Young human beings, like all

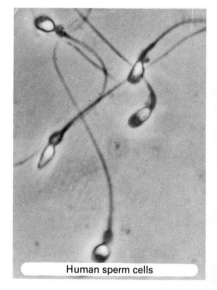

Human sperm cells

young mammals, are not able to reproduce. Their reproductive organs are reduced in size and they do not produce sex cells. The changes which cause a child to become an adult are referred to as puberty. They begin when the pituitary gland, on the underside

Principles of Biology

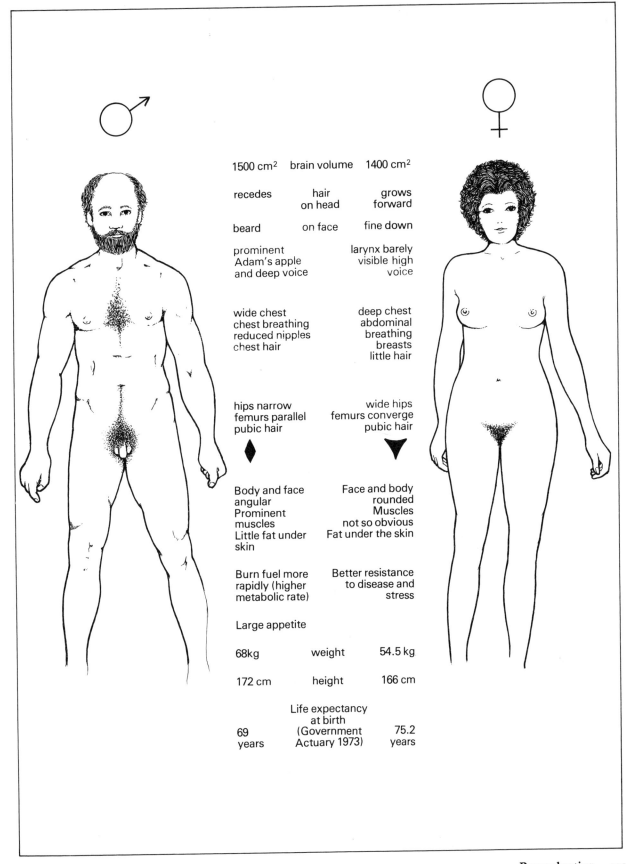

of the brain, produces hormones which directly affect the ovaries or testes. These begin to produce sex cells and also their own hormones, chemicals which bring about all the physical changes involved in becoming a man or woman.

The onset of puberty in girls is marked by the beginning of the menstrual cycle (periods). An egg begins to ripen in the ovary. At the same time a thick lining to the uterus develops. About fourteen days later the egg breaks out of the ovary and starts its journey to the uterus. Two weeks after the egg is released, if it has not been fertilized, the lining of the uterus breaks down and passes out of the body with some blood. This is called menstruation and it continues for a few days. The whole cycle then begins again with the ripening of another egg, and it will continue until the woman is about 55 years old, apart from the times when she is pregnant. The cycle is controlled by hormones. For example, a hormone is released into the blood when an egg is shed from the ovary. It prevents further egg-release until the previous one has either passed out at menstruation or completed its development into a new individual.

In most mammals, patterns of reproductive behaviour are fixed and fairly predictable; man seems to have greater freedom and therefore less predictable behaviour patterns. These are acquired either by 'trial and error' or are taught by parents or other members of the group. Most human societies have developed methods of regulating and controlling this sexual freedom. There are marriage customs, usually with quite harsh rules preventing, for example, marriages between closely-related people. Marriage caters for the fact that all human beings need continuing bonds of affection with others and also that the young take many years to mature. Sexual activity has to be pleasant, otherwise people would not take part in it and the species would not survive. The rules in most societies make sure that human beings act responsibly in their sexual activities and do not merely use each other for pleasure.

Reproduction in other mammals

In most mammals the two oviducts each merge into a uterus, the two uteri meeting at the beginning of the vagina. They also produce several offspring at a time. Mammals show a regular cycle of sexual activity. Some of them (pigs, rabbits, mice and rats) have an inbuilt rhythm. Others breed only at certain times in the year, perhaps only once a year, depending on temperature, or the availability of food. Sheep, for example, mate in the autumn and produce their lambs in spring. Solitary animals, like wild hamsters, meet only at mating time, the female producing a scent which attracts the male to her. Animals living in herds, such as grazing animals or seals, are dominated by a single powerful male who drives off all the young males. They remain bachelors except for the one who becomes strong enough to defeat the bull in combat and take over control of the females and the young.

The development of the young mammal at birth varies a great deal. Young born in protected nests are often blind and helpless at birth. The female giraffe gives birth to her calf standing up and it is able to run about from birth. Most mammals do not menstruate. When the female ovulates (sheds the egg from the ovaries) she is said to be in heat and will actively seek the male. If by chance she does not become pregnant, the lining of the uterus is reabsorbed, not shed as it is in human beings.

Growth

Growth follows a similar pattern for all living things. It occurs when active cell-division takes place. It can be measured either as increase in size or as increase in weight. The graph shows the increase in weight of a green plant over a period of time. It would have a similar shape if it were a graph of animal growth. The increase in weight of a population of bacteria would change in the same way. It can be seen that the *rate* of growth increases to a maximum. This follows from the fact that the more cells there are, the more cells there are to divide. Growth slows as maturity is reached. It then falls off as the tissues die and lose water. In human beings, loss of weight occurs as muscles waste away. Some people continue to gain weight but this is usually because they eat too much.

Mammals change their proportions as they get bigger. This is because different parts of the body grow at different rates. Some plants have leaves whose shape depends on how big they are.

There are important differences between growth in plants and in animals. Plants grow only in particular regions, called meristems. These are at the tips of roots and shoots. Trees and shrubs increase in girth because of the activity of the cambium, a thin sheath of actively-dividing cells outside the wood and immediately below the bark. It produces cells towards the inside and these become wood (xylem). It also produces cells towards the outside

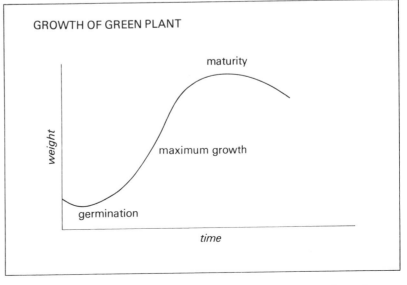

and these develop into phloem, which is replaced each year. Animals grow all over, by division of cells in all parts except nervous tissue – no division of nerve cells occurs, only increase in the length of nerve fibres.

Plants continue to grow until they die. Mammals grow to a certain size and then growth stops. In man, cell division still occurs in the skin, and where blood cells are formed, but this is for replacement purposes only.

Plant growth is seasonal, whereas mammals grow continuously to full size provided they have adequate food.

Growth in plants is affected by all sorts of conditions – climate, rainfall, fertility of the soil, competition from other plants, situation. Mammals generally grow faster and bigger the more food they have, although it is likely that they die sooner.

In both plants and animals, chemical substances affect growth. In plants, these substances are plant hormones (auxins); in animals, they are the hormones produced by the pituitary and thyroid glands.

Cheese plant showing change in form from young to mature leaves

Heredity

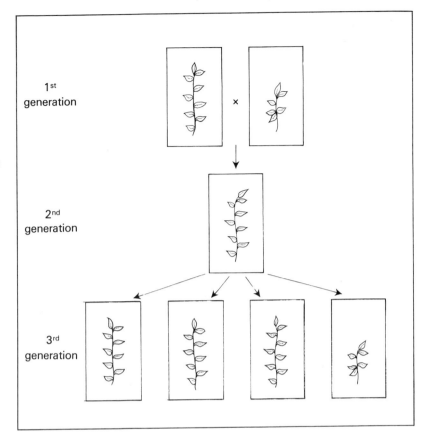

The study of heredity seeks answers to such questions as: why do children resemble their parents? Why do they differ from them and from their brothers and sisters? Why do some people have light skins while others are dark?

No two people are exactly alike. They can be divided into two groups according to sex; they can be separated quite independently into four groups on the basis of their blood groups; for characteristics such as height it is not possible to divide them into groups at all. People range from the very tall to the very short with the majority somewhere about the average. Even in the smallest group – perhaps a pair of identical twins – people may look alike, but they are not in fact identical.

There are two main ways of answering questions about heredity. We may carry out controlled breeding experiments on suitable plants and animals, or we may study the structure and behaviour of cells.

Man has been successfully breeding varieties of domestic animals and plants for centuries, but it was only just over a hundred years ago that scientific study of heredity began. It was carried out by a monk, Gregor Mendel, living in that part of Europe now known as Czechoslovakia. He was successful because he simplified the problems. Instead of asking difficult general questions like those we started with, he was content to ask much more limited questions. He used the garden pea for his experiments. This exists in a few clearly distinct varieties and is normally self-pollinated. Most pea plants grow to a height of about 1.5 m but there is a dwarf variety about 0.5 m tall. Peas are never exactly the same height (height is affected by other factors such as temperature, situation, and food supply) but it is not possible to mistake tall pea plants for the dwarf variety.

One of Mendel's experiments was designed to see what happened when tall plants were cross-pollinated with dwarf plants. The anthers from the flowers of tall plants were removed before they were ripe and their stigmas dusted with pollen from dwarf plants. The reverse experiment was also done.

The seeds from his plants were collected and sown in the following spring. All his plants this time were tall. He allowed them to self-pollinate. Again he collected the seeds. The following year about three-quarters of his plants were tall and the rest dwarf. The same experiment has been repeated many times. As the results are added together, the ratio of tall to dwarf approaches the exact proportion 3:1.

Similar patterns of inheritance are known for man. In this case it is not, of course, possible to carry out breeding experiments. We have to look for breeding patterns by searching the records of family histories which show the patterns that interest us.

A man from a family with a history of brown eyes, and no other colour, married to a blue-eyed woman, has children all of whom are brown-eyed. The same is true if the mother has brown eyes and the father blue eyes. Marriages between such brown-eyed people, i.e. those whose parents had different eye-colours, produce approximately three times as many brown-eyed as blue-eyed children.

In both these examples, one of the characteristics disappears in the second generation, reappearing in the third. This is not always so. When snapdragons with red flowers are crossed with white-flowered snapdragons, all the next generation are pink-flowered. When the pink generation interbreeds, it produces red, pink and white-flowered plants in the proportions 1:2:1.

The following terms are commonly used in heredity:
A **pure-breeding** plant is one which produces only tall plants like itself when self-pollinated. 'Pure-breeding brown eye' means that two such people can produce only brown-eyed children.

Characteristics like tallness in peas, or brown eyes in man, are said to be **dominant**. Features such as dwarfness in peas, or blue eyes in man, are called **recessive**. Parents hand on something to their offspring which is passed on from one generation to the next (although it may not show its effects in every generation).

The factors which are handed on are called **genes**.

Cell division
Cell division starts with division of the nucleus. It appears to break up, after a number of strands or rods become visible. These are **chromosomes**, and their number is constant for all the body cells of all members of a species. Man, for example, has forty-six chromosomes. The number is always even because every recognisably-different chromosome is present twice, i.e. there are two complete sets.

The first stage in cell division is the doubling of all the chromosomes so that each of the two new

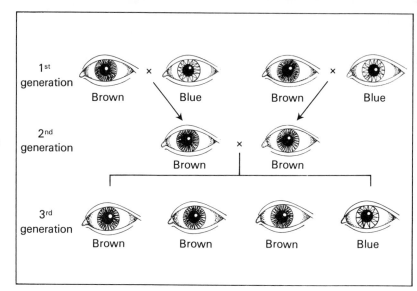

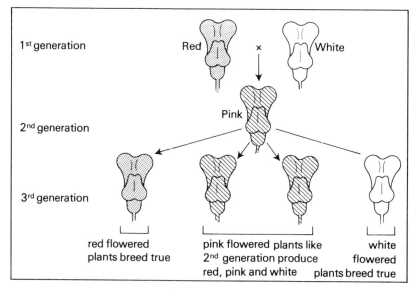

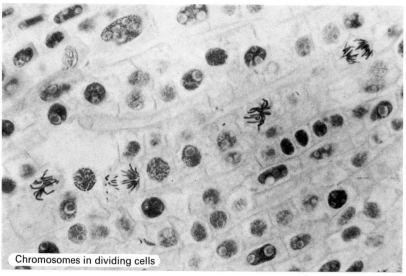

Chromosomes in dividing cells

cells receives an exact copy of every chromosome present in the parent cell. Since each pair of chromosomes behaves in the same way during cell division, we can show in diagram form what happens to one pair.

When sex cells are being formed by division, the chromosomes behave differently. Each sex cell, male and female, has only half the number of chromosomes found in ordinary body cells. Only one member of each chromosome pair is present, i.e. one complete set. When fertilization takes place, the full number is restored when the sperm cell (male) fuses with the egg cell (female). Each pair of chromosomes in all our body cells is a copy of an original pair (one from father and one from mother) present in the fertilized egg.

This is what happens to one pair of chromosomes in the pea-breeding experiment.

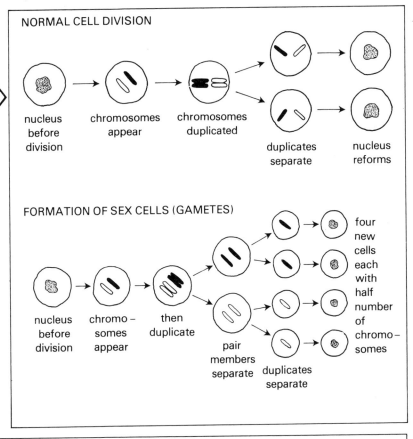

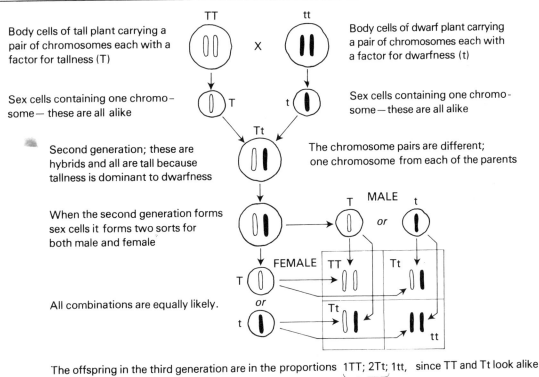

Heredity 137

In snapdragons, where there is no dominance, the hybrids can be identified because they look different. It is not possible to tell by inspection of the other examples which is hybrid and which is pure. We can find out only by breeding. A brown-eyed hybrid married to a blue-eyed person has, on average, children of both eye colours in equal proportions. A hybrid tall plant crossed with the dwarf pea plant produces both tall and dwarf offspring in equal numbers.

Chromosomes are known to be made up of long chains of genes arranged in order like beads on a string. The total number of genes in a cell amounts probably to many thousands. They control all cell behaviour and activity. They are like a set of instructions which is handed on from one generation to the next. Because the genes are in pairs, which are not always alike (T or t), and because they are organised in a number of separate groups, the chromosomes, it means that during formation of sex cells new combinations are possible. In this respect genes are like a pack of playing cards. During the production of sex cells, the whole deck is reshuffled and a new hand dealt to the new individual.

Sex in human beings, as in all mammals, is explained in a different way. Women have a pair of chromosomes known as X chromosomes included among their twenty-three pairs. Men have only one X chromosome; its partner, which is different and much smaller, is called the Y chromosome. Each egg cell will contain an X chromosome, but the sperm cells will be of two sorts, depending on whether they contain an X or a Y chromosome. If the egg is fertilized by an X-carrying sperm, it will develop into a girl. If it is fertilized by a Y-carrying sperm, it will develop into a boy.

X-chromosomes carry genes, but as far as we know there are no genes on the Y chromosome. A recessive gene, causing colour-blindness, is found on the X chromosome. A man with this gene on his X chromosome will be unable to distinguish between greens and reds. All his sons will have normal vision because they do not receive an X chromosome from him. Provided his wife is normal and has no family history of colour-blindness, his daughters will also be normal even though they get the recessive gene from their father; this is because their normal mother will have provided the normal dominant gene. The daughters in their turn will transmit X chromosomes carrying the recessive gene to both their sons and daughters. Half the children will receive an X chromosome carrying the normal gene, and some of their sons will be colourblind like their maternal grandfather.

Characteristics inherited in this way are sex-linked. A man who shows the characteristic transmits it through his daughters, who do not show it, to his grandsons. Colourblindness, like all sex-linked characteristics, is rare in women.

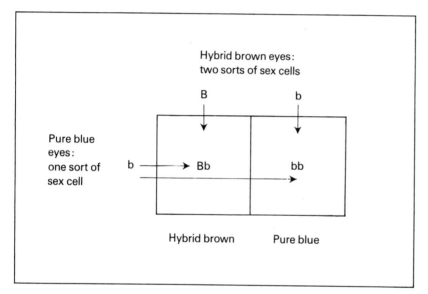

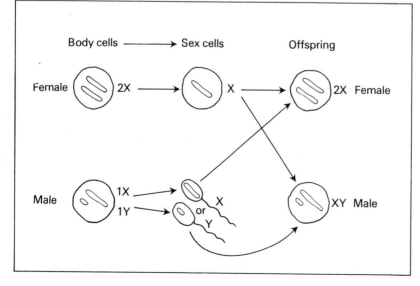

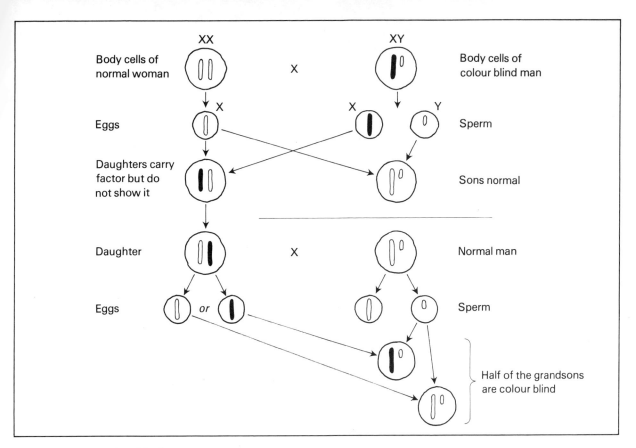

Women who are colourblind must have a colourblind father and a mother who also carries the gene.

Heredity

Evolution

Fossil remains of *Ammonites* in a slab of limestone (A),
Mammoth that has been preserved with its meat and skin due to eternal frost (B)

The theory of evolution is one of the most important in biology. It is important because it accounts for so many facts which cannot be explained in any other way. It accounts for both the differences and resemblances between living organisms, satisfactorily explains the meaning of fossils, and provides a history of life on earth.

The theory states that living things gradually and continuously change with the passage of time and that these changes are handed on from generation to generation. It means that organisms living in the distant past must have been very different from those living today. It means that all organisms now living are related to one another. The more alike they are, the closer the relationship between them must be.

The case for evolution rests on evidence drawn from three different sources.

First, there can be little doubt that fossils are the remains, or evidence of the remains, of living things. The ages of the rocks in which they are found can be determined with considerable accuracy. The further back into the past we go, the more the fossils differ from modern organisms. Fossil evidence strongly suggests that life was different and that it has changed although it cannot prove that all living things it represents were related to each other.

Second, the classification of plants and animals is based on their structure, similarities, and differences. When drawn as a plan it looks like an enormous family tree. This makes sense *only if it is a family tree*.

Third, we know that offspring are never exactly the same as their parents. We cannot know that this has always been so but it is reasonable to suppose that it has. Although such differences produce little effect over the course of centuries, they become very important during the millions of years evolution has been taking place.

Variation is present in every species. No two individuals are ever exactly alike, not even identical twins. No two individuals ever carry exactly the same collection of genes.

Genes sometimes get altered. Such an alteration is called a **mutation**. Mutations are probably caused by naturally-occurring radiation particles. It is known that artificially-produced radiation (X-rays) causes mutations. When mutations occur in developing germ cells, they may be transmitted to the next generation and thus continually add to the variation in the species. Most organisms at the start of their lives have a very small chance of surviving to the adult stage and reproducing in their turn. Most will die by accident, but although the chances of survival are small they are not the same for all. Some will have a better chance than others. This automatic survival of the fittest is called **natural selection**.

No living organism exists in a vacuum. It lives in, and is part of, its **environment.** No two environments are ever exactly alike. The needs for survival will differ in each. In this way, the environment 'selects' the organisms best fitted to survive in it. The environment of intense tropical sunshine has selected men with dark, rather than light, skins.

These four factors – **variation**, continually added to by **mutation**, subject to **natural selection** by the **environment** – ensure that a slow process of change continuously occurs. Because it is so slow, the balance of nature is not upset, and it is only when we are able to look back over the vast stretches of time revealed by the fossil record that we can know that there has been any change at all.

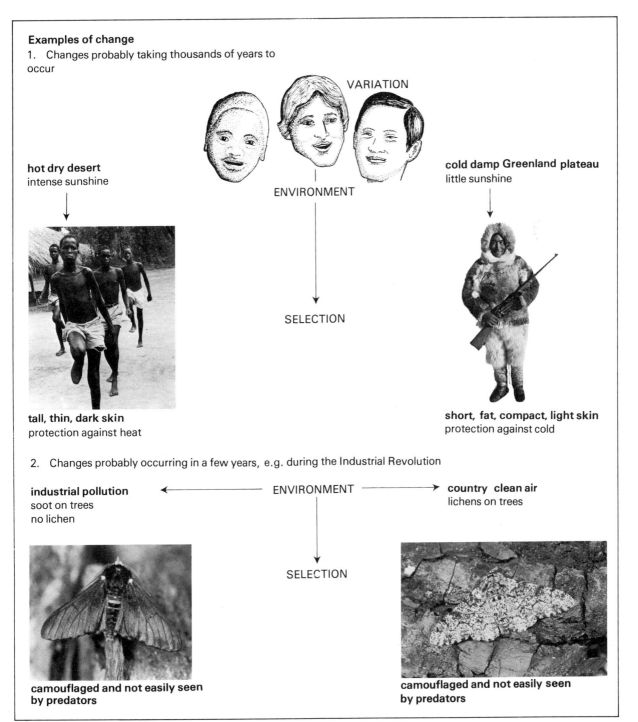

Examples of change
1. Changes probably taking thousands of years to occur

VARIATION

ENVIRONMENT

hot dry desert
intense sunshine

cold damp Greenland plateau
little sunshine

SELECTION

tall, thin, dark skin
protection against heat

short, fat, compact, light skin
protection against cold

2. Changes probably occurring in a few years, e.g. during the Industrial Revolution

industrial pollution ← ENVIRONMENT → **country clean air**
soot on trees
no lichen

lichens on trees

SELECTION

camouflaged and not easily seen by predators

camouflaged and not easily seen by predators

Evolution 141

Time in millions of years

1

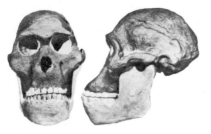
Model of restored skull of early man

Early man (*Australopithecus*)

150

Fossil remains of a primitive bird, *Archaeopteryx*

A primitive bird— *Archaeopteryx*

200

Fossil remains of *Diplodocus*

An ancient reptile— *Diplodocus*

370

Fossil remains of an early fish *Cephalaspis*

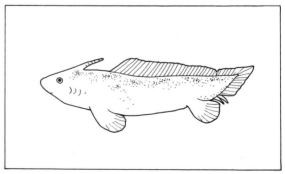
Early fish (*Cephalaspis*)

142 **Principles of Biology**

Part 5
The community

The community

Living organisms do not exist in isolation from each other. Although they may live apart they depend for their existence upon each other, just as living cells do in a single plant or animal.

It is possible to study man as a single living being. It is also possible to study men in groups, in the family, in the village, in the city or in a country. We may study transport, communications, trade or commerce. In the same way it is possible to study plants and animals living together and affecting each other.

The study of living things in their natural surroundings, and the effects they have on each other, is called **ecology**.

Some other important words you should know are:
Biosphere – the whole living world
Ecosystem – a natural region and all the living things in it, together with the non-biological features which influence it
Community – all the living things in an ecosystem
Niche – the 'home' of a single type of living organism in the community.

The biosphere

The biosphere is the largest unit of life. It contains and includes every living organism. If the whole of the earth were suddenly to disappear, except for living things, we would be left with a structure looking rather like a giant soap bubble. Its outer surface lies somewhere in the outer layers of the atmosphere. Its inner surface would be a few feet below the surface of the land and perhaps as much as a mile below the surface of the oceans.

All the materials from which living things are made are found in the biosphere. They circulate continuously. The total amount remains constant. Nothing is added, nothing is taken away. The biosphere is like an enormous machine. Energy is needed to drive it. The source of its energy is the sun. Nothing else is added to the system from outside.

The biosphere contains large amounts of water and the chemical element carbon. Without these there would be no biosphere. Water has a number of important properties. It will dissolve more materials than any other liquid. As ice, it is lighter than its liquid form. Volume for volume it holds more heat than almost any other material. Carbon atoms can be chemically

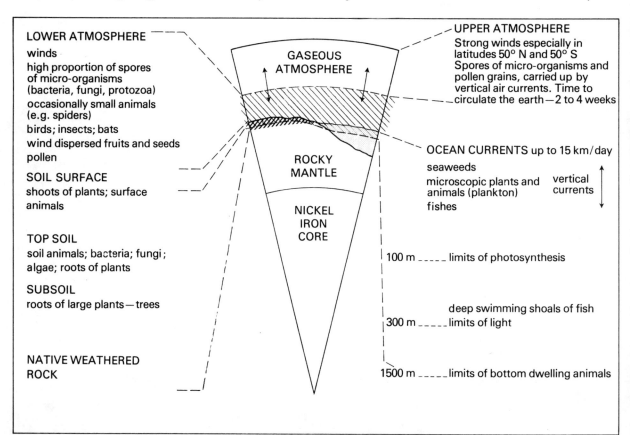

144 **Principles of Biology**

combined together in very large numbers. This makes possible the great variety of complex substances that are needed for life. No other element is known to have this property.

The raw materials which are needed for life are:

Carbon dioxide – a gas in the atmosphere or dissolved in water
Oxygen – a gas in the atmosphere or dissolved in water
Water – most of which is in the oceans
Chemical elements (in addition to carbon, hydrogen and oxygen); generally combined as simple salts in solution in water. Some of the more important ones are: nitrogen, iron, magnesium, calcium, potassium, sodium, phosphorus and sulphur.

These materials are built up into the complex substances, mainly carbohydrates, fats and proteins, which make up the tissues of plants and animals. This is a continuous process. At the same time raw materials are constantly being made available again as organisms die and are decomposed. Energy is needed for this re-cycling of materials. The whole cycle of life can be thought of in some ways as a giant water wheel.

The circulation of some materials essential to life is shown below.

Water is present in abundance. The oceans provide a home for large numbers of plants and

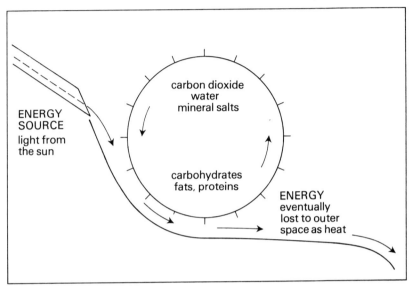

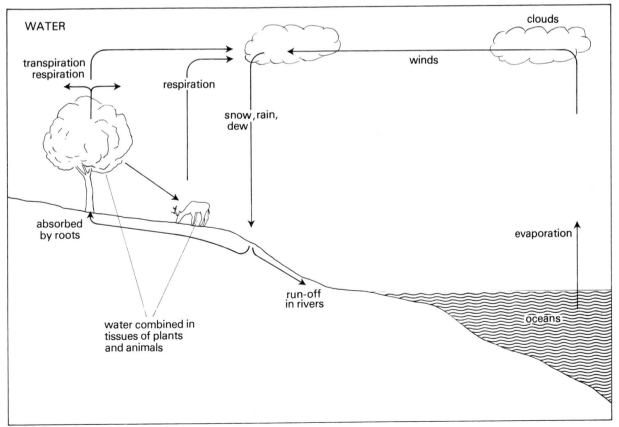

The community 145

animals, and an inexhaustible supply of water for land plants and animals when the water evaporated from the sea falls as rain on the land. Sunlight provides the energy which drives this particular cycle. Without it there could be no life on dry land.

Carbon is an essential part of all carbohydrates, fats and proteins.

Carbon dioxide forms about 0·04 per cent of the earth's atmosphere. It is estimated that this would be used up in about ten days by the process of photosynthesis. In fact, the amount of carbon dioxide used up in photosynthesis is exactly balanced by the amount of carbon dioxide produced by respiration. The level of carbon dioxide in the atmosphere could be noticeably altered only if the rate of burning coal and oil were enormously increased and if the forests of the earth were considerably reduced.

Nitrogen is an essential part of proteins. Most living organisms are unable to use the nitrogen gas which forms about 80 per cent of the earth's atmosphere. Much of the nitrate in the soil gets washed down into the deeper layers by rain and eventually finds its way into the sea. This loss of nitrate is probably balanced by the activity of the nitrogen-fixing bacteria.

Phosphorus is an important part of cells, found combined as nucleic acids in the nuclei of plant and animal cells.

Plants obtain phosphorus in the form of phosphate in solution in the soil. The supply of phosphate is added to by the slow breakdown (weathering) of the rocks. Some of the phosphate is lost, like nitrates, by draining into the seas, finally ending up on the ocean floor. Phosphorus is therefore a 'key' element. The total amount of living matter on the earth at any time is determined by how much phosphorus is available as phosphate.

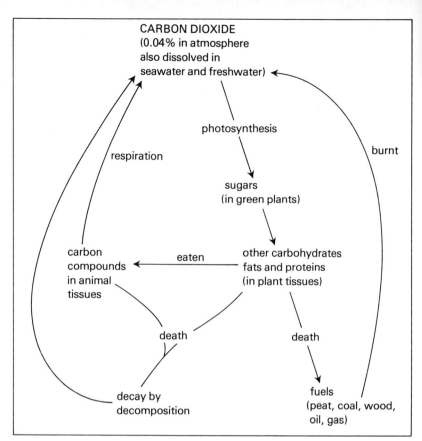

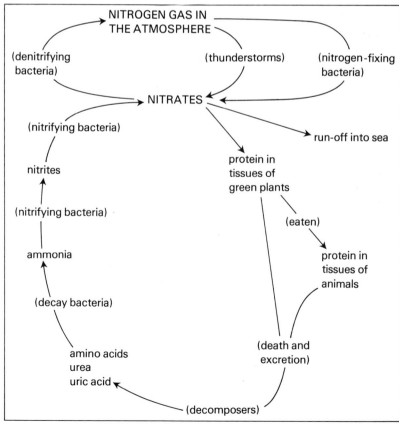

If the loss to the ocean floor is more than the amount released by weathering of rocks, the total amount of living organisms must decrease, simply because the weathering of the rocks cannot be speeded up. Man is able to meet his needs for phosphate for agricultural fertilizers by obtaining it from guano, fossilized bird droppings. This is not unlimited in quantity, and neither is it replaceable.

Energy

Energy is the only ingredient which is added from outside the biosphere. The Law of Conservation of Energy states that energy cannot be destroyed. A piece of wood contains energy. When we burn it in a fire the chemical energy it contains is changed into heat energy. This is gradually dispersed and so is no longer available to do work. Water in a reservoir high up in the mountains has energy because of its position. When it is allowed to run downhill (as in hydro-electric schemes), this energy is converted into electrical energy. Whatever use we make of that electrical energy, it eventually becomes changed into heat energy. It is still energy but it cannot be used to do work and is lost to the biosphere. The fate of the energy reaching the earth from the sun is shown in outline, right.

The diagram on p. 148 shows that the energy that reaches the earth from the sun eventually re-radiates into space as heat. The earth loses as much energy as it gains. Some of this energy passes through living organisms. Their tissues therefore contain energy. We eat food in order to acquire that energy and it is released by the process of respiration. It is also released when we burn plant tissues, such as wood or paper, coal or oil. The energy value of plant and animal tissue is measured in terms of the heat it

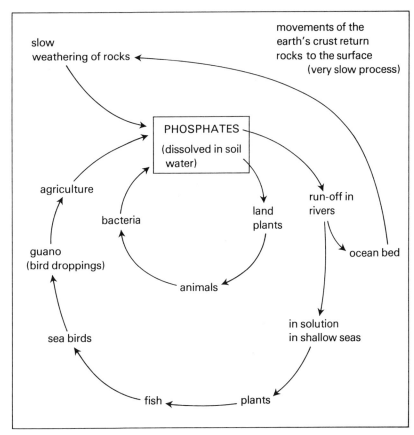

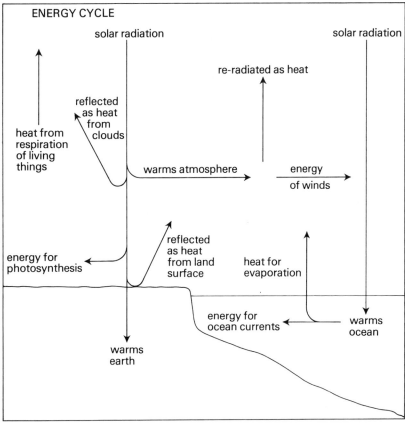

The community 147

produces when it is burnt. It is expressed as **kilocalories per gram**. A kilocalorie is the amount of heat needed to raise the temperature of 1 000 cubic centimetres of water by one degree centigrade. One gram of fat produces about 9 kilocalories, carbohydrate and protein about 4½ kilocalories. With this information we can now construct an 'energy ladder'. Suppose we start with 1 000 kilocalories of energy produced by the green leaves of a plant (this will be a little over 200 g of sugar), see below.

It will be seen that at each step some of the energy absorbed in the form of food is used in the vital activities of the organism. In a mammal which feeds on grass, much of its energy intake is used for movement and maintaining body temperature. A large part of its food is never absorbed but is eliminated as faeces. Only about 10 per cent of its total food intake actually adds to its body weight. Plants are more efficient in changing energy into body tissue since they do not use up energy in moving about.

If we now replace the steps in the energy ladder by the names of actual living organisms we have what biologists call a food chain. Most food chains contain only a few organisms. This is due to the great loss that occurs as energy is transferred from one step to the next, i.e. as one animal gets eaten by the next in the chain.

The following are two examples:

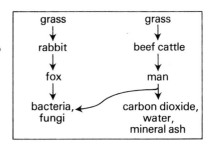

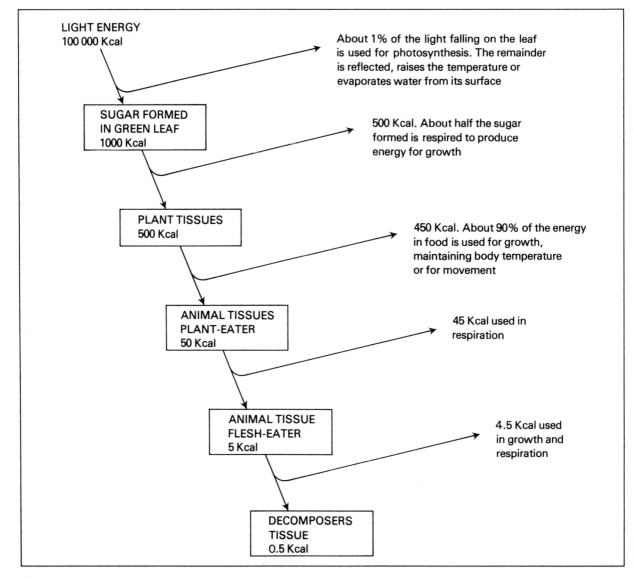

148   **Principles of Biology**

If we now replace energy value (expressed in kilocalories) by weight, it is not too difficult to see that the total weight of living organisms at each step in the food chain drops off very rapidly. For example, if a field produces 1 000 kilograms of grass per year, this could be turned into not more than 100 kg of beef cattle. This amount of beef, eaten by a group of schoolchildren, will not add more than 10 kg to their combined weight. Food chains can normally contain only a few steps because of this rapid fall-off in transferred energy. Biologists describe this decrease as a pyramid of numbers (see right).

A pyramid of numbers implies that as we climb the pyramid the animals or organisms become bigger. This is not necessarily the case. A dog may support a large population of fleas without suffering much. The important point is that it is the **weight** of organisms that decreases at each stage. Food chains and pyramids of numbers are, in any case, rare. The relations between each rung of the energy ladder are best represented by a food web.

The diagram below presents the general features of a food web. There are basically three ways that material can pass through it. The most obvious one is that which includes all the larger animals. They all eat live food. Scavengers have been included. These are animals like the hyena, vulture and carrion crow. Their food is probably dead when they start to eat it, but then this is sometimes true of the flesh-eaters. This is the living food chain.

The second is called a detritus food chain. It is based on the dead remains of plants and animals. Although the final breakdown of tissue is carried out by bacteria, a whole variety of plants and animals maintain themselves in this chain.

A third route lies through the parasites. Strictly speaking parasites are really a special form of the living food chain. It so happens that they are surrounded by their food (for example, worms in a dog or pig). Note: parasites usually get eaten when their host gets eaten. For example, a duck eating a snail

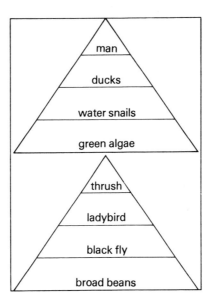

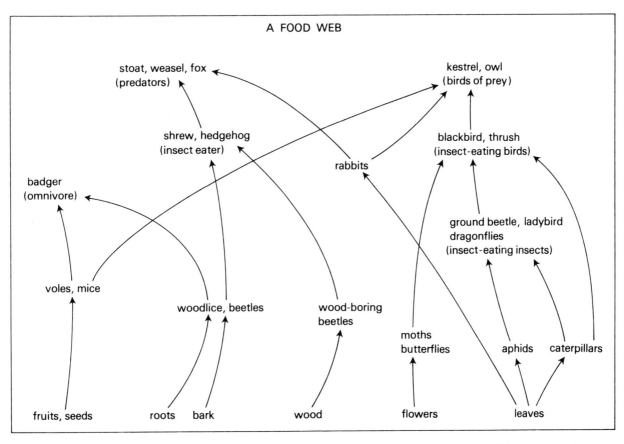

The community 149

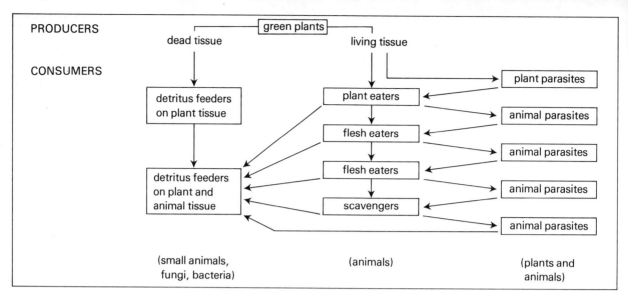

digests its parasites as well. This is not always the case. Parasites sometimes transfer their activities to a new host, for example larva of tapeworm in a rabbit may become adult tapeworm in the dog that eats the rabbit.

There are a number of terms describing the members of a food chain. These are:

Producer organisms – these are green plants. Their basic requirements are carbon dioxide, water and a supply of mineral salts. They make sugar by photosynthesis and they can build up proteins.

Consumer organisms – all other living organisms, with the exception of some of the bacteria. These can neither make sugars from carbon dioxide and water, nor can they make protein. They obtain these materials ready-made from green plants, either directly or indirectly, dead or alive.

Herbivores – all animals which feed only on plant tissue.

Carnivores – animals which obtain their food by eating other animals.

Scavengers – these are animals which feed on the 'crumbs from the table' (examples: jackal, hyena, vulture, carrion crow).

Parasites – animals or plants which use other animals or plants as a home as well as a food supply. They are always smaller than their hosts who get no advantage from the arrangement (indeed they may suffer harm). The parasite gets food and shelter.

Saprophytes – fungi and some bacteria which get their food from dead organisms.

Other special relationships:

Symbiosis – this refers to organisms of different species which live in close association with each other to their mutual advantage. Neither can survive without the other.
Example: some termites feed on wood. They are not able to digest it but the single-celled animals (protozoa) which live in their digestive systems can. The termite swallows the wood, the protozoa digest it, both live on it.

Commensalism – this describes a close relationship in which each partner gains, although the relationship is not essential to survival. Example: a hermit crab living in a discarded shell often carries a sea-anemone around with it. The crab gets protection, the sea anemone gets transport.

Ecosystems

The biosphere displays great variety. Much of it in Britain is artificial because man has organised it to suit his own purposes: meadows, ploughed fields, hopfields, orchards, gardens and so on. But there are many parts which have not been subject to man's activity, such as the small patches of the original oak forests which covered much of England. Many parts of the seashore remain unspoiled; as do many of the ponds, lakes and rivers. On a larger scale, the variety in the biosphere is illustrated by the Arctic tundra with its sparse covering of mosses and lichens, the tropical rainforest of the Amazon basin, the deserts and some tropical islands. They are all different because of variations in such factors as rainfall and temperature. Any region which appears to be separate and distinct and can be studied independently is called an **ecosystem**. It may be any size, from a rainwater butt to half a continent. Its study includes not only the living things in it but also the non-living factors which affect it. Since plants are at the base of any food chain, they determine what consumer organisms are present.

These factors govern the types of plant present on land:

temperature – both average temperature and range of temperature
rainfall – both total amount and how it is delivered
soil – structure and composition
light – both intensity and duration
winds – both force and direction
height above sea level

and in the sea:
temperature
depth
light
currents, affecting distribution of mineral salts.

Rainfall: for large land ecosystems, rainfall determines whether they are desert, grassland, woodland or forest.

Temperature: in the case of forests, temperature decides whether they are coniferous (in cold climates), deciduous (in temperate regions) or tropical.

Soil: in temperate deciduous forest, the soil type decides whether it is oak (clay soils), ash (limestone soils) or beech (chalk soils).

Oak forests are fairly open and admit lots of light. There are many smaller trees, shrubs, and ground plants. Beech woods are dark because their leaves exclude light. When the leaves fall, they form a waterproof layer on the forest floor and this discourages other plants from growing.

An oasis in the North African desert (A),
Savannah – tropical grassland (B),
Tropical forest – mangrove swamp (C),
Plantation of conifers (Douglas fir) – these trees cast deep shade and very little grows at ground level (D),
Oakwood – light easily penetrates to ground level and there is plenty of plant cover (E),
Beechwood in Sussex – little light reaches ground level and plant cover is sparse (F),
An ash forest (G, overleaf)

The community 151

G

A   B

Ecosystems can be thought of in the same way as business organisations. How efficient are they at producing the goods? How good are ecosystems at forming tissue or using solar energy? The diagram below compares a number of ecosystems from this point of view.

It may seem strange that the shallow seas of a coral reef can be as productive as forest or grassland. Forests contain masses of plant tissue whereas a bucket of sea water seems not to contain any plants at all. Forests seem to be composed mainly of plants but in the sea, animals are most conspicuous.

The difference is partly explained by the fact that forests contain a lot of permanent tissue which is hardly ever eaten. On the other hand, plant material in the seas is eaten almost as quickly as it is formed. The rates at which tissue is formed and transformed can be the same in each case. The difference can be illustrated in the following way:

A This man has an income of £50 per week. He spends £50 per week. He has £1000 in the bank. (The £1000 in the bank is like the permanent forest tissue – hardly ever touched.)

B This man has an income of £50 per week. He spends £50 per week. He pays his bills on payday. He never has more than £50 at a time and then not for long. (This man's pay is like plant material in the seas – used straight away.)

Soil

The soil is an ecosystem. Each kind of soil provides a different ecosystem. Man has been learning about soil, through his mistakes, ever since he changed from a hunter or a fruit gatherer to a cultivator of land. There is still a lot to be discovered about soil and its care.

Although the soil has a life of its own, it is important in other ecosystems for the following reasons:
1   It provides anchorage for green plants
2   It is the source of the water needed by plants
3   Its atmosphere is a source of the oxygen needed for the underground parts of plants
4   It provides the mineral salts in solution necessary for the growth of plants
5   It contains the micro-organisms which re-cycle materials by causing decay.

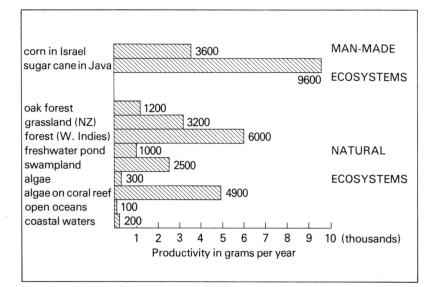

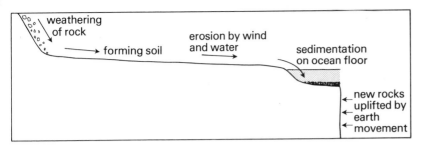

The origins of soil

Soil is continuously being formed. It would not be inaccurate to say that there is a soil cycle.

Soil is formed from parent rock by a process called weathering. This may be caused by:

1 Changes in temperature causing unequal expansion and contraction. The rocks therefore crack.
2 Rainwater lying in the cracks expands as it freezes, thus splitting the rocks into smaller pieces.
3 Moving water, as rain, as streams, or as ice in glaciers wears away the rock and carries away the particles.
4 Dissolved carbon dioxide in rainwater may act as a very weak acid dissolving away the rock, especially limestone and chalk. These are examples of physical weathering. Biological weathering also occurs. Lichens and mosses on bare rock gradually break down its surface. Their remains mix with the rock particles to provide a favourable growing situation for other plants. Their roots enter crevices and split the rock still further. A crack in a concrete path will soon become wider if the weeds that grow in it are not removed.

The structure of soil

Soil contains:

1 Particles of various sizes, resulting from the breakdown of rock. These are chiefly sand from the weathering of quartz, and clay (resulting from the breakdown of aluminium silicates).

2 Organic matter. This consists of the breakdown products of plants and animals, dead soil organisms, fungal threads and the smallest remains of plant tissues. This is called humus; it gives a good soil its dark colour.
3 A vast population of living organisms. These include bacteria, fungi, microscopic green plants, protozoa, roundworms, earthworms, small arthropods, together with the seeds and underground parts of flowering plants.
4 A substantial amount of air containing rather more carbon dioxide than the air above.
5 Water as a thin film surrounding the soil particles.
6 Mineral salts in solution.

The analysis of soil

A If some soil is placed in a glass jar, a quantity of water is added and the whole mixture is well

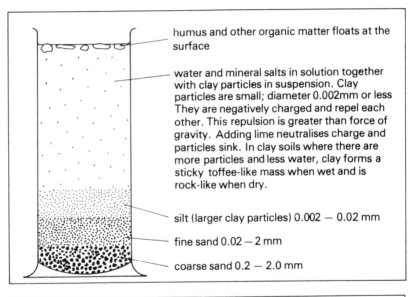

A SOIL PROFILE

Top soil: usually about 20 to 30 cm deep.
It contains small, weathered particles and is usually dark because of the organic matter (humus) in it formed from the breakdown of plants and animals

Subsoil: contains larger, partly-weathered particles.
It is usually much lighter because it does not contain humus.

The community 153

shaken (cover the open end first), the different soil parts will separate as shown on page 153.

B To determine the percentage of water in the soil:
1 Weigh an evaporating dish.
2 Add a sample of soil and weigh again. Subtract weight of dish to obtain weight of soil.
3 Place sample in a drying oven at a temperature slightly below 100°C.
4 Re-weigh at intervals of a day or two until there is no further weight loss. Cool the dish before weighing.

Result: $\dfrac{\text{loss of weight}}{\text{original weight of soil}} \times 100$

= percentage of water in soil

C To determine the percentage of humus in the soil:
1 Weigh a crucible.
2 Add sample of soil from experiment **B** and weigh again. Subtract weight of crucible to obtain weight of sample.
3 Heat strongly to red heat.
4 Cool and re-weigh.

Result: $\dfrac{\text{loss of weight}}{\text{original weight}} \times 100$

= percentage of humus in soil

D To determine the percentage of rock particles in soil:
1 Weigh a crucible.
2 Add a sample of fresh soil and then proceed as in **C2, 3, 4**.

Result: $\dfrac{\text{final weight of soil}}{\text{weight of soil before heating}} \times 100$

= percentage of rock particles in soil

E To determine the percentage of air in the soil:
1 Place suitable tin in glass container and fill with water as shown. Mark water level (a).

2 Remove tin, allow water on its outside to run back into glass container (b).
3 Empty tin and find its volume by filling from a measuring cylinder (c).
4 Punch holes in the bottom of the tin.
5 Drive tin into soil as shown.
6 Dig out the tin, removing surplus soil from top and sides.
7 Place tin in the glass container and, with a glass rod, poke out the soil to release the air.
8 Add water from a measuring cylinder to bring the water back to its original level (d).

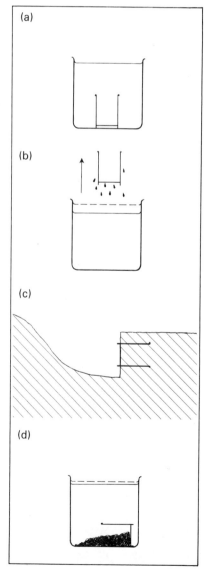

Result: $\dfrac{\text{amount of water added at }\mathbf{8}}{\text{volume of tin}} \times 100$

= percentage volume of air in soil

F To demonstrate the presence of micro-organisms in the soil:

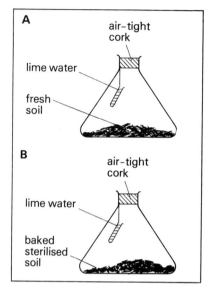

1 Set up the apparatus and leave for a few days.
Result: the limewater in flask A turns milky showing that carbon dioxide is given off as a result of bacterial respiration. The micro-organisms in flask B are dead.

G To compare the capillarity (ability to draw up water from deeper layers) of sand, clay and loam soils.
1 Set up the apparatus as shown. All the glass tubes are plugged at the bottom with cotton wool. Make sure the three soils are dry to begin with and pack them equally tightly.
2 Fill glass trough almost to the top with water.
3 Watch upward movement of water in tubes for several minutes after immersion. Watch at regular intervals (e.g. daily) until no further rise is visible.

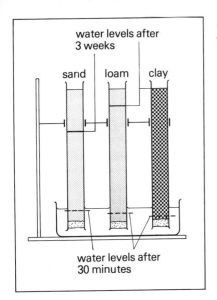

Result: water passes quickly into the sand to fill the large air spaces. Final water level is highest in clay because of its greater capillarity.

H To compare the drainage rates of different soils:
1 Set up the apparatus shown below. Each funnel is plugged with cotton wool. Take care to see that you use equal volumes of soil, packed equally tightly.
2 To each funnel, at the same time, add 50 cm³ of water.
3 After the water ceases to flow, note the results.

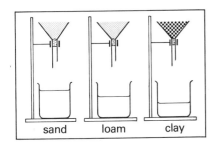

The texture of soil
The structure of soil depends largely on the proportions of sand and clay it contains and on the amount of humus present. **Sandy soils** containing little humus tend to be dry soils. They are well aerated but have little capacity for holding water.

Clay soils tend to be poorly aerated because the particles are so small. They tend to hold water and do not drain easily. In dry weather, the clay forms lumps which are almost rock-like. In wet weather, it is sticky and hard to separate. **Loam soils** have a mixture of clay and sand particles. Soils of this kind have a definite structure. The particles collect together to form 'crumbs' The presence of humus in the soil improves the crumb structure. In sandy soils, it improves its water-holding capacity; in clay soils, aeration is improved.

Some agricultural practices
Spring ploughing – By increasing surface area, it speeds up the drying-out of the soil. It aerates the soil, encouraging the activity of bacteria.
Autumn ploughing – Organic matter is ploughed in and decomposes more quickly. Frost can penetrate the soil and helps to break down the large lumps.
Hoeing – Aerates the soil, reduces water loss and exposes and damages the roots of weed plants.
Rolling – Compresses the surface layers of soil, bringing more capillary water to the surface. This helps dry the soil or increase the rate of seed germination.
Crop rotation – The replacement of one crop plant by another which makes different demands on the soil. Clover, beans or peas may be grown and then ploughed in. These crops, called legumes, add to the nitrate in the soil because their roots contain bacteria which are able to convert the nitrogen gas in the air into nitrate.

Methods of preventing soil erosion
1 **Preventing erosion by water**
Terracing the land breaks up the slopes. Ploughing along the contours also reduces the run-off of water which might wash away the soil. Planting trees on slopes helps to hold the soil in position.
2 **Preventing erosion by wind**
Hedges dividing the land act as windbreaks. Trees may be planted to act as shelter belts. Planting cover crops instead of leaving the land fallow, and strip-cropping (planting grass or legumes between strips of grain crops), also help hold the soil in place.

Other ecosystems
Apart from forming a complete ecosystem with its own great variety of living organisms, the soil provides the basis for all the other land ecosystems. In addition to physical factors such as rainfall and temperature, soil composition has an important effect on the character of ecosystems. Three ecosystems are described below. In two of them, the freshwater pond and oakwood, the nature of the soil is an important factor.

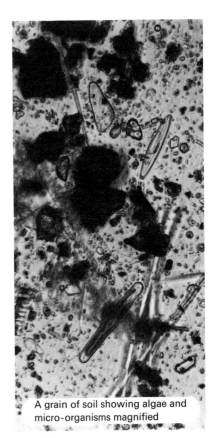

A grain of soil showing algae and micro-organisms magnified

The community

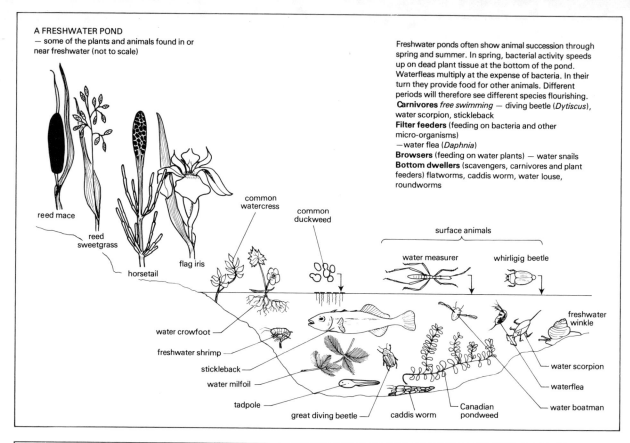

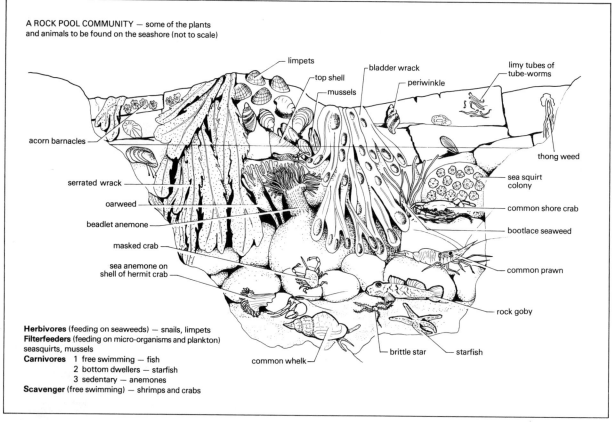

The growth of ecosystems

Ecosystems grow and change just as animals and plants do. If no outside circumstances interfere, they reach a 'grown up' stage and then change no more. As everything depends on the plant life, this growth and development can best be illustrated by reference to plant life. Wherever bare soil is exposed, for example on motorway embankments or in sandpits, the plants begin to move in. Sometimes these are algae, lichens or mosses, especially if there is little soil and a lot of exposed rock surface. Then the annuals move in, plants which live for one year only. Gradually, they create conditions more favourable to perennial plants, which live for a number of years. As time passes, shrubs appear and then, finally, trees. There are many examples of this process going on all over Britain: pieces of land such as demolition sites, old factory and railway areas no longer used, derelict buildings and gardens.

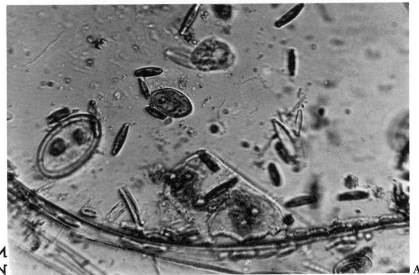

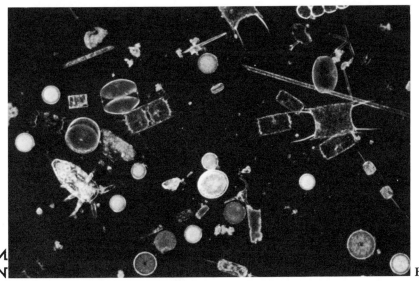

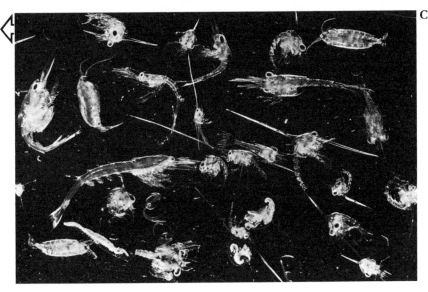

Magnification of various micro-organisms in a drop of pond water (A), Microscopic plants living in the surface layer of the ocean (phytoplankton) (B), Some of the small animals living in the surface layers of the ocean (zooplankton) Most of these are crustaceans (C)

The community 157

An oakwood community

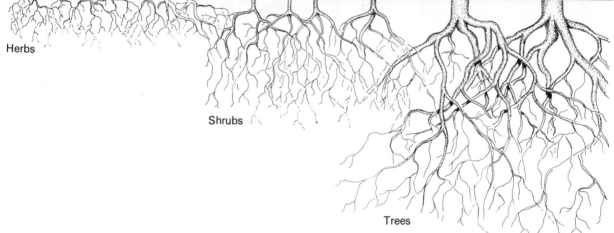

Herbs

Shrubs

Trees

Other trees

hazel

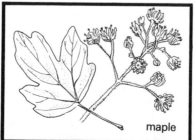

maple

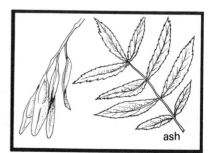

ash

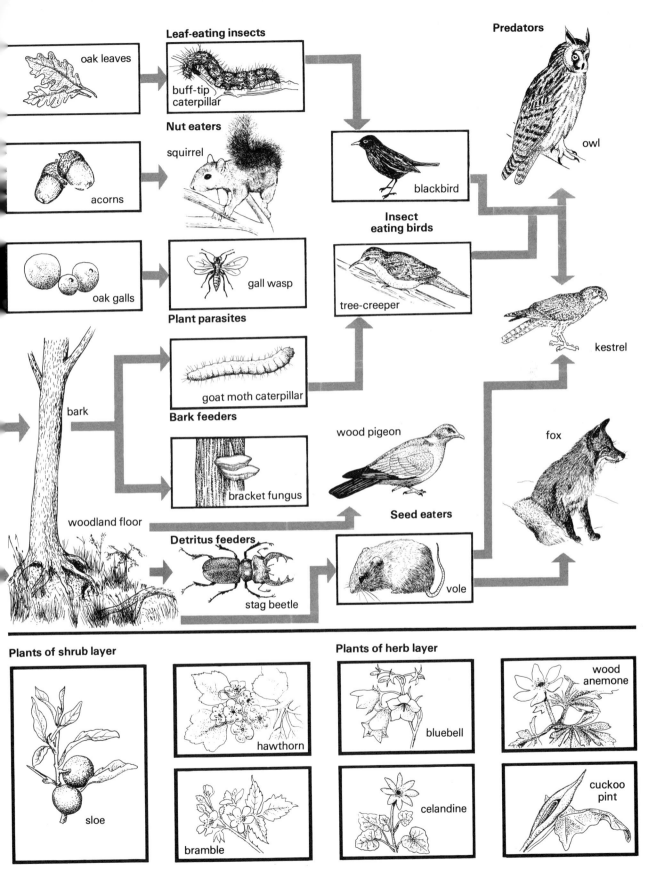

The community 159

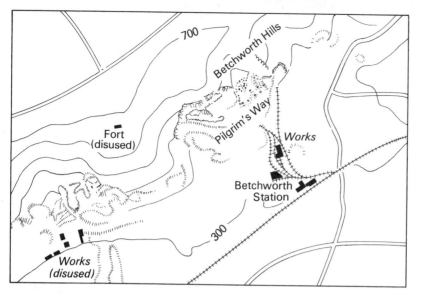

The map shows a section of the North Downs in Surrey. Quarrying operations have gradually shifted eastward. The most westerly part of the quarry is therefore oldest. The photographs show the process of plant succession on chalkland over a period of perhaps fifty years.

Nearly a hundred years ago, a piece of ground at the Rothamsted Experimental Station in Hertfordshire was fenced off and left without interference from man. It has now become an oak wood, not very different from the few natural oak woods which still exist.

Many ecosystems are man-made and have to be worked on to make sure that they stay as they are. When man ceases to look after them, they change. Here are some examples:

Betchworth Quarry, Surrey: exposed chalk visible – working recently discontinued (A),
Another part of the quarry ten years since quarrying has ceased – hawthorn scrub and wild roses dominate the vegetation (B),
An older part of the quarry where quarrying ceased about thirty years ago – dominant plants are birch spindle and especially yew (C),
Broadbalk wilderness at Rothamsted (D),
Stretch of typical downland (E),
Hawthorn scrub growing on downland no longer grazed (F),
Typical heathland (G),
Birchwood (H)

A

B

C

Chalk downland
Downland owes its character to generations of grazing by sheep and by rabbits. Close grazing prevents the seedlings of trees, and other woody perennials from becoming established. When sheep no longer graze and the rabbits disappear, the appearance of the landscape changes.

Heathland
Heathland is maintained by grazing and by periodic burning. When this no longer occurs, it may then turn into woodland – oak woodland on

The community 161

clay soils and birch on sandy soils.

Ponds

Ponds are sometimes maintained by feeder streams. If they are not cleaned out from time to time, ponds may silt up and completely change their character.

Regulation of numbers of plants and animals
Factors affecting the numbers of plants

Three important factors are light, temperature, and water supply. Increasing the amount of **light** will increase the amount of growth, up to a certain point. A rise in **temperature** without altering the amount of light will also increase the growth rate. If these two factors remain constant, then increasing the amount of **water** available to plants will increase their growth rate.

In most situations, one or other of these factors will act as a limiting factor for growth. Temperature and light tend to go together. The nearer to the poles, the less vegetation tends to grow. Tropical forests are obviously more productive than Siberian tundra. Similarly, productivity varies with the availability of water.

The nature of the tree cover can also affect plant growth and numbers. Beech trees have shallow roots. Their leaves stop a lot of light from reaching ground level. Their dead leaves form a dense layer on the ground. The soil tends to be dry underneath. An oak wood allows more light to penetrate to the ground. Its roots are found at deeper levels. The dead leaves curl up as they dry and do not cover the ground so closely.

The speed at which the various mineral salts are passed through the food chains will also affect productivity. In tropical forests, decomposition is rapid and the mineral salts pass back into the forest vegetation very quickly. In acid or waterlogged soils, like wet heathland, decomposition is slow and plant growth correspondingly slow.

Population control in animals

In natural situations, plant-eating animals are never found without flesh-eaters also being present. There have been enough instances to show that when flesh-eaters are not present, the plant-eaters increase their numbers rapidly, over-graze the food supply, and die in large numbers from lack of food. They may even cause erosion of soil by stripping the plant cover. The intro-

A typical pond near Woldringham, Surrey

The site of what was once a pond near Ottery St Mary, Devon. Silt, carried into it by its feeder stream has been allowed to collect. The pond gradually changed into a bog thicket containing alder and willow as well as irises, bulrushes, sedges and reeds.

duction of the rabbit into Australia and the Scottish red deer into parts of New Zealand are examples. In its new home, neither of these animals had a natural enemy and each has increased its population greatly, causing damage to plant life.

Plant-eaters produce lots of offspring. Their numbers are controlled by their predators and, in times of scarcity, by food shortage. Their numbers are, at any particular time, fixed *by the number surviving*. The number of flesh-eaters in any particular community is determined by the number of their prey. As the population of plant-eaters rises with increasing food supplies, so does the population of flesh-eaters. As plant food decreases, so does the population of plant-eaters. This is closely followed by a decrease in the number of flesh-eaters. The levels of both populations are closely linked.

Flesh-eaters play an important part in controlling the numbers of plant-eaters, but their own numbers are controlled in a more complicated way. They do not generally act as a food source themselves, so their numbers can only be related to the availability of their prey animals. It seems that it is their breeding rates which are affected by the availability of food. In times of scarcity they breed less, in times of plenty they breed more. Their numbers are at any particular time, fixed by *their rate of breeding*.

How this operates is only partly understood. Some species show what is called territorial behaviour, in common with many fish and with birds. The male drives off other males from his territory by his threatening behaviour (they seldom actually fight). The scarcer food becomes the more aggressive he becomes and the larger his territory. The result is that some males are excluded altogether and fail to acquire a mate and to start a family. The population becomes more thinly spread over the ground. In times of plenty the opposite happens.

Birds which live in large groups show pecking orders, in which strong birds dominate. The more difficult it becomes to find food, the more stress there is in the group. As a result, more individuals are pushed to the lowest part of the pecking order and do not breed.

In normal circumstances, plant-eaters will never eat all the available plant food. The flesh-eaters will see that they do not. By linking rate of reproduction to availability of food, flesh-eaters make sure that they also never eat all their food supply.

The place of man

There have been creatures called men on the earth for thousands of years, if not millions. Only during the last few thousand years has man had a history, in the sense that he has brought about changes in his way of life.

Pre-Stone-Age man was probably just as clever as modern man, but with no language and little knowledge, it takes a long time to change things, however clever man may be. Once a language has evolved and can be written down, progress is rapid. All experience can be shared, passed on to following generations, and accumulated.

At first, man was probably a food-gatherer, living on fruits and any other available plant material. Australian aborigines in the outback still live this way. With the development of crude weapons, man became a hunter and widened the range of his diet to include meat. His total numbers were small and he would still have been a natural part of his ecosystems, just like any other animal. From late Stone Age times until a few hundred years ago he was a tiller of the soil, lived in settled communities and domesticated animals as well as plants for his own use. During this time, damage was done to his environment. Much of the land around the shores of the Mediterranean Sea became desert through over-grazing by the sheep and goats of wandering tribesmen. During the sixteenth century, large areas of oak forest were felled in Britain to build ships for the navy. Areas of the mid-western United States became a dustbowl because men did not know how to care for the land. In lowland Britain, most of the natural environment was transformed into the pastures and meadows of today.

The third phase of man's history began about two hundred years ago. It began with a massive increase in population, and was rapidly followed by the Industrial Revolution which provided jobs in new industries. Indeed, had there been no industrial revolution many of these people would have died through lack of food. As it turned out, new industries provided the means of support for many extra people as well as helping still further to increase their numbers. Land was taken for building factories, mills, railways, and extending towns and cities. The overall effects of the Industrial Revolution, as we can now see, are:

1 Massive destruction of natural habitats and species, both animal and plant.
2 Extensive pollution of the environment.
3 Loss of natural resources, many of which cannot be replaced.

Instead of taking his place in a balanced system, man ignores the system and turns the materials of the natural world into his material possessions.

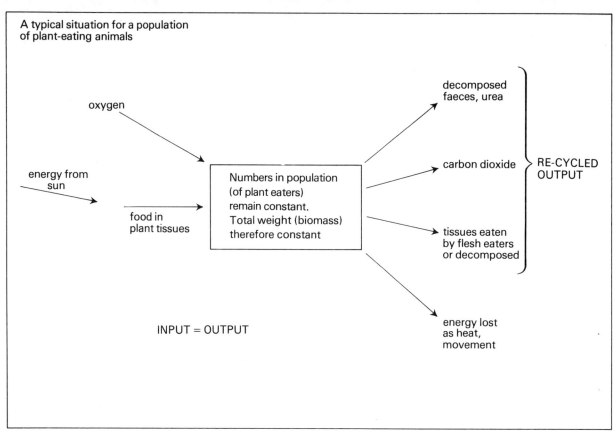
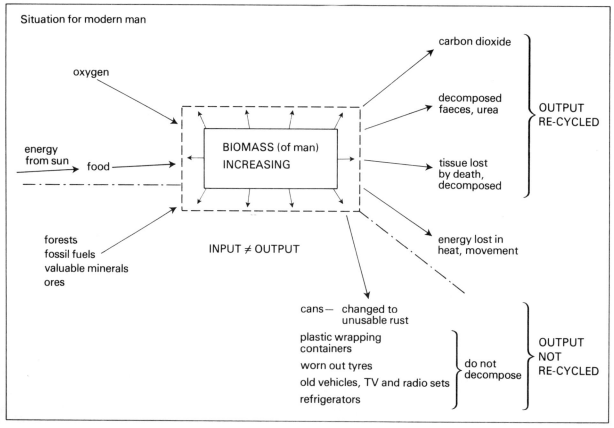

164 Principles of Biology

Man's effects on his environment	Action to remedy
Increasing population	Legal abortion, the pill, the loop, sterilisation } methods of preventing births or reducing birth rate
Increasing sewage	Development of better sewage treatment works; use of energy from decomposing sewage; return of remaining solids to the land
Destroying natural ecosystems to grow more food	No remedy
Increasing pollution of the sea	International agreements to prevent oil dumping, dispersal of radioactive material and nerve gases
. . . of the land	Laws enforced to restrict dumping, the use of some pesticides and over-use of nitrates
. . . of freshwater	River boards set up to control discharge of sewage and factory waste
. . . of the air	Laws to govern discharge into the air from industrial sites, research into the control of exhaust gases
Litter dumping	Fines for litter louts (difficult to enforce), campaigns, e.g. Keep Britain Tidy
Depletion of natural resources Renewable	Plant-a-Tree campaigns (present paper consumption exceeds rate at which world's trees are growing) Laws to prevent over-fishing
Non-renewable	Active search for alternative energy sources such as solar batteries
Shortage of land	High density building

The table above is neither complete nor completely accurate. You may feel that there are really only two important items on the left side – the increasing population, and the desire for more and more of the 'good' things in life like clothes, furniture and houses, steak and chips, foreign holidays, air travel and personal transport. All other items in the list seem to result from these two.

The remedies on the right side seem to be of two kinds also. Laws, restriction, controls, fines, all make life complicated and uncomfortable. Who can possibly know all the laws? How can they all ever be enforced? The other remedies also make life more complicated. They demand more responsibility and 'know-how' from everybody, for example in birth control. They demand more know-how from society about pollution and its effects, the development of new energy sources, and the long-term effects of accumulating and getting rid of radioactive waste.

There is no doubt that as problems become more difficult, so the chances of solving them become less. The biosphere and its ecosystems are much older than man. It and they will eventually come to a new balance point. The big question is, will it be an equilibrium achieved by the co-operation and common consent of all mankind, through social or political action, based on an understanding of the problems? Or will it be left to nature?

Things to do

1   Collect photographs, in colour if possible, showing different sorts of ecosystem, natural or man-made.
2   Look for and record examples of good conservation and its opposite.
3   Try to arrange to dig a small piece of ground and then keep a record of the changes that occur in it.
4   Take each of the examples in the table. Use each as a basis for some investigation of your own. For example, what steps are being taken by your own local council to deal with rubbish disposal? How much waste heat is lost from its power stations? What is it doing to preserve worthwhile countryside? Does it operate in ways which are harmful to plant and animal life – spraying verges with weedkiller, for example?

Anti-conservation

166　**Principles of Biology**

Index

absorption 80, 81, 90–91, 93, 102
accommodation, eye 117
adaptation 36, 42, 48, 49
adrenalin 117
aerobic respiration 95
agricultural practices 155
algae 54–5
   filamentous 55
alimentary canal 76, 81
alligators 40
amino-acids 3, 4, 83, 100
Amoeba 11, 112
amphibians 38–9
anaerobic respiration 95
annuals 69
antibodies 87, 88
anti-toxins 87
Antrodemus 40
Arachnida 26
arteries 86, 88, 101
Arthropods 26–7
assimilation 81
atmosphere 2, 3, 73, 144
auricle 89
auxin 122

backbone 106, 107, 110
bacteria 50–52, 80, 87
bat 49
beer-making 58
bees 20
biennials 69
bile 80, 82
bile duct 82
binary fission 10
biosphere 144–5
birds 42–5
Birkenia 33
bivalves 28
blackbirds 43
bladder 103
blood 32, 82, 83, 84–9, 99, 100, 103, 105
blood groups 85, 135

blood pressure 102
blood vessels 84, 86, 101, 107, 129
bone 106–108
Bowman's capsule 101, 102
bracket fungi 57
brain 49, 99, 114, 115, 119
bread-making 58
breathing 95, 97–9
Brontosaurus 40
buds 67
bulbs 68
butterflies 22

caecum 80
capillaries 85, 86
carbohydrates 3, 4, 75, 76
carbon 144, 146
carbon compounds 3, 4
carbon dioxide 96, 98, 99, 146
carnivores 150
cartilage 107
caterpillars 22
cell division 10, 66, 133, 136–7
cells 6–7, 13, 83, 85, 94, 97, 104, 106, 112
cellulose 80
centipedes 27
chameleon 41
Chlamydomonas 54
chloroplast 55
chordates 32
chromosomes 136–9
chyme 79
cilia 10
circulation 84, 88
clubmosses 62
cochlea 119, 120
coelenterates 12–13
colonial animals 12
colour-blindness 138, 139
commensalism 150
community 144
computers 115
conifers 64–5
cortex 67
crabs 26
crocodiles 40
Crustacea 26
cuttlefish 29
cytoplasm 6

dehydration 51
desert rat 49

diaphragm 98
diatoms 55
diets 76
diffusion 7, 97
digestion 79
disease 50, 87
dogs 78
dominant characteristics 136
downland 161
dragonfly 23
duck-billed platypus 46
duodenum 79, 80

ear 119–21
earthworms 17, 18, 117
echinoderms 30–31
ecology 144
ecosystems 150, 152, 155
eels 37
effector cells 113
egestion 81, 100
eggs 42, 129, 130, 137
endoskeleton 106
energy 2, 72, 75, 76, 83, 95, 100, 102, 144, 147, 149
energy ladder 149
environment 141, 165
enzymes 74, 76, 80, 81
epidermis 66, 67, 69, 104
Erysiphe 57
Euglena 10
evolution 140–41
excretion 6, 100
exoskeleton 19, 106
eye 117–19

faeces 83
fats 3, 4, 75, 76
femur 107
fermentation 96, 97
ferns 62, 63
fertilization 124, 137
fibrin 88
fibrinogen 88
fish 34–6
flatworms 14–15
flesh-eaters 162, 163
flukes 15
fluorine 75
foetus 129
food 72
food chain 149, 150
food preservation 51

food substances 75
food web 149
forests 151, 152
fossils 140
freshwater pond 156
frogs 38, 39, 47, 48
fungi 56–8

gastric juice 79
Gastropods 28
genes 136, 139, 140
gerbil 49
germination 127
gibbons 49
gills 58
glands 105, 116, 117
globe fish 36
glycogen 83
gnats 22
gravity, sensitivity to 119
growth 6, 133, 162

hair 105
head movement, sensitivity to 119
hearing 119
heart 84, 88–90
heartbeats 90
heathland 161
herbivores 150
herbs 67
heredity 135
honey bees 20
honeycomb 20
hormones 116, 117, 130
horsetails 62
housefly 22
hydra 12, 13
hyphae 56, 57

ileum 80
immunity 88
Industrial Revolution 163
inflammation 87
insects 19, 23–5, 106
insulin 117
intestinal juice 80
iodine 75
iron 75
irritability 112

jellyfish 12
joints 108, 109

kangaroo 46, 47

kidneys 100–103
kilocalorie 149
koala bears 46, 47

larynx 97
leaf-fall 92
leaf forms 68
leeches 17
lichens 59
life, raw materials for 145
ligaments 107
liver 82
liver fluke 15
liverworts 60, 61
lizards 41
lobsters 26
locusts 21
lugworms 16
lungfish 37
lungs 98, 99
lymph 87, 88
lymph vessel 85

magnesium 75
mammals 46–9
  adaptation 48, 49
  growth 133, 134
  nutrition 74
  placental 47
  pouched 46
  primitive 46
  reproduction 130
marsupials 46, 47
Mendel, Gregor 135
menstrual cycle 130
micro-organisms 157
mildews 57
millipedes 27
mineral salts 75
mites 26–7
mole 48
molluscs 28–9
monkey 49
monotremes 46
mosses 60, 61
motor nerves 115
moulds 56, 57
Mucor 56
muscles 79, 90, 95, 106–109
mushrooms 57, 58
mussel 28–9
mutation 140
myriapods 27

nastic reactions 121
natural selection 140–41
nerve cells 113, 114
nerve fibre 115, 119
nervous system 32, 84, 114, 116
newts 38
nitrogen 75, 146
notochord 32
nucleus 6, 7
nutrition 6, 72–83

oakwood 160
octopuses 29
osmoregulators 106
osmosis 7, 91–3
ostriches 45
oxygen 73, 96–9

pancreas 117
pancreatic juice 80
Paramecium 10
parasites 14, 149, 150
Pasteur, Louis 51, 52
pasteurization 52
pea plants 135
Pelycosaur 40
penguins 45
Penicillium 57
pepsin 79
perennials 69
pericardium 89
peristalsis 79
phloem 66, 67
phosphorus 75, 146
photosynthesis 63, 72, 75, 100
placenta 47, 48, 129, 130
planarians 14
plant-eaters 162–3, 164
plants 4, 6, 60, 64, 75, 84, 95, 99, 100, 112, 121, 129, 157, 159
  deciduous 92
  effect of gravity on stems and roots 122
  effect of light on stems 122
  flowering 66–9, 90
  green 150
  growth 133, 134, 162
  herbaceous 67
  pure-breeding 136
  reproduction 123–8
  woody 67
plasma 87
platelets 87

Pleurococcus 54
polar bear 49
pollination 124
pollution 163, 165
Polytrichum 61
ponds 162
population 162, 165
potometer 93
prawns 26
predators 159
proteins 3, 4, 75, 76, 100
Protozoa 10
pseudopodium 11
puberty 130
puffballs 57

rabbits 77, 80, 81, 110
radiation 52
ragworms 16
rays 36
recessive characteristics 136
red corpuscles 87
reflex action 116
refrigeration 51
rennin 79
reproduction 6, 123
    asexual 128
    human 129–32
    in other mammals 130
    plants 123–8
    sexual 123, 129–30
    vegetative 128
reptiles 40–41
respiration 6, 10, 95–7, 149
respiratory system 99
retina 117, 118
rock pool community 156
roots and root-hairs 66, 90–91, 93

salamanders 38
saprophytes 150
scavengers 150
scorpions 26–7
sea anemone 12
sea urchins 30
seal 49
seaweed 54–5
secretin 80
seed development 128
seed dispersal 126–7
seed growth 126–7
sense organs 117
sensory cells 113, 119

sex cells 130, 137, 139
sexual intercourse 129–30
sexual reproduction 123
shark 34
sheep 78
shells 28, 29, 106
shrimps 26
single-celled animals 10–11, 112
single-celled organisms 50
skates 36
skeleton 106–111
skin 104–105
slipped disc 110
slugs 28
snails 28, 196
snakes 41
soil erosion 155
soil profile 153
soil 4, 18, 50, 151
    analysis 153
    as ecosystem 152
    formation 153
    origins of 153
    structure 153
    texture 155
sole 36
spiders 26–7
spiny anteater 46
Spirogyra 55
squids 29
starches 3
starfish 30–31
sterilization 51
stomach 79
sugars 3, 83, 91
sulphur 75
sun 2
swallowing 78
sweat 105
symbiosis 59, 81, 150

tadpoles 38, 39
tapeworms 14, 15
Tasmanian wolf 47
teeth 76–8
    carnassial 78, 79
temperature 151
tendons 107
ticks 26–7
tissue fluid 85
toads 39
toadstools 57
tortoises 40, 41

trachea 97
translocation 94
transpiration 91–3
transport
    blood, in mammals 84–9
    water, in plants 89–94
trees 158, 162
tropisms 121
trout 35, 36
tubeworms 16
turtles 40

ureter 101
urethra 130
urine 103
uterus 130

valves 86–90
variation in species 140
veins 82, 86, 89
ventricle 89
vertebrates 32–3
villus 80, 81, 83
viruses 53
vitamins 75
Vorticella 10

waste removal 81, 100
water 2–3, 102, 144, 145
water cultures 74
water fleas 26
water snail 15
water vapour 93
whale 49
white corpuscles 87
windpipe 97
wine-making 58
winter twig 67
woodlice 26
worms 16–18

xylem 66, 67, 91, 92, 94

yeast 58, 97